CW00589580

Wild Plants of Glasgow

Wild Plants of Glasgow

Conservation in the City
and Countryside

J H Dickson

Paintings by Elspeth Harrigan
Photography by T N Tait

ABERDEEN UNIVERSITY PRESS
1991

Dedication

To the members of the Glasgow Natural History Society

First published 1991
Aberdeen University Press

The publisher acknowledges subsidy from the Scottish Arts Council towards the publication of this volume.

British Library in Publication Data
Wild plants of Glasgow: conservation in the city and countryside
I. Dickson, J. H.
582.130941443

ISBN 0 08 041200 9

Design by Mark Blackadder
Typeset by Hewer Text Composition Services, Edinburgh
Printed by Bookcraft (Bath) Limited

Foreword

by David Bellamy

Glasgow, not quite the place to inspire a budding botanist you might think, but you'd be wrong.

The environs of Glasgow, complete with its tenements, parks, bings, coups and the waxing and waning of sunset and sunrise industries, have been an inspiration to generations of field botanists from David Ure to the author of this fascinating little book.

A book which tells us in detail that much of Glasgow's rich natural plant heritage is still intact, wooing future Jim Dicksons to follow in his footsteps. One hundred plants have gone to the wall of local extinction and some are still under threat, but in these days of green enlightenment their safety lies in expert knowledge and common understanding, with which this book abounds.

In those not so far off pre-motor car days every town worth its salt had a natural history society and Glasgow's, though not the oldest, was always held in high esteem as was and is the Department of Botany at the university, complete with Regius Chair. It is the painstaking work and vision of the members of both these institutions, past and present, which form the inspiration of these pages. A proud record of caring people who know the importance and excitement of a local patch of

Primroses or Water Horsetail and so put their finger on the pulse of this key part of Scotland's natural heritage, ***the wild plants of Glasgow***.

I hope that this book will become an integral part of the library of every Glaswegian household. A book of beauty and great charm which will lead more and more people into the ways of conservation, at the same time allowing the ecesis of future generations of botanists.

On my next trip to Glasgow, at the right time of year, I am certainly going to get local expert help so I too can take a look at Young's, or is it Glasgow's, own Helleborine Orchid.

Save the wild flowers and we will save the world.

David Bellamy
Bedburn 1991

Preface

As a youth and undergraduate when I often consulted and annotated John Lee's *Flora of the Clyde Area* I longed for a book about local plants that was not just informative but attractive. Later when I came to use the Floras by Roger Hennedy and Thomas Hopkirk these works were also very useful but just as uninteresting in appearance as John Lee's book. The colourfulness and low price of the present book have been made possible by the generous subsidies provided by Shanks and McEwan plc, The Nature Conservancy Council for Scotland, The Incorporation of Gardeners, The University of Glasgow and The New Phytologist Trust. I hope that young people will be attracted into studying and conserving the many plants that grow in and around Glasgow.

I am especially appreciative of the skills of my two collaborators whose results grace these pages. A friend of many years, Norman Tait is a dedicated and highly experienced natural history photographer. A friend of only a few years, Elspeth Harrigan has a talent given to very few: this is the ability to combine scientifically accurate depiction with a very pleasing artistry.

Several members of the Glasgow Natural History Society as well as some non-members are mentioned by name in the text because they made discoveries of the plants discussed and illustrated in the book. Many other members of the society and some non-members contributed records including very note-

because they made discoveries of the plants discussed and illustrated in the book. Many other members of the society and some non-members contributed records including very noteworthy ones that are not considered here. All these many helpers will be acknowledged in the full Flora of Glasgow when it is published. To everyone I am most grateful for their companionship, diligence and keen-eyed powers of observation.

The portraits in Chapter 3 are reproduced by permission of the Glasgow Natural History Society.

To my wife Camilla I owe a great deal, not just for her skill in drawing the maps but for all her help over the years.

JHD

Contents

List of Plates

List of Figures

PART 1

Introduction

CHAPTER 1

Background

Forty years have passed since, as a secondary schoolboy living in Ibrox, I began to hunt for plants. Only a few minutes' walk away was Bellahouston Park, at the northern tip of a green corridor penetrating the city from the south-west. Pollok Estate to the south, now Pollok Country Park, the largest of the many public parks in Glasgow, was not much further away. At that time, Glasgow could still perhaps be thought of as the Second City of the Empire. The heavy industry persisted, if not flourishing as it had in late Victorian times. Shipbuilding, locomotive works, coal mining and steelworks were still operating within and around the city. A badly polluted Clyde and its tributaries flowed through a Glasgow bustling with crowded tramcars. The people, some consumptive and some bandy-legged as a consequence of rickets, for much of the year breathed air heavily laden with sulphur dioxide and smoke from many thousands of chimneys with coal-burning fires, both domestic and industrial. Smogs did not just stop the traffic but killed some of those inhabitants with already weakened lungs or hearts.

Now, in 1991, we live in the post-industrial age or the New Environmental Age as it has been called by the noted conservationist, Max Nicholson. The city, its environment and people have greatly changed, in some ways much for the better, not least in the greatly decreased smoke and sulphur dioxide. Just as the Environmental Health Department of the City of Glasgow District Council monitors the air pollution, so the Clyde River Purification Board checks the now greatly improved quality of the river and its tributaries. In 1983 salmon returned to the Clyde after an absence of 120 years. However, as pointed out by Desmond Hammerton, Director of the Clyde River Purification Board, buried chemical wastes from Glasgow's uncontrolled industrial past remain a threat to the rivers.

Some 200 years ago the first men to search for plants around Glasgow and publish lists of their findings were the hard-up Rev David Ure and the well-to-do Thomas Hopkirk. In the last decade of the eighteenth century the size of Glasgow was still such that stacks of peat for fuel and haystacks for fodder could be found in the Trongate at the very heart of the city. The Hopkirks' *country* house was at Dalbeth (near Tollcross) then well outside the city, but soon to be enclosed to the north and west by housing and the east and south by heavy industry.

Fifty years later, during the time of Roger Hennedy, that keen populariser of botany, Glasgow had greatly developed, although there was much expansion still to come. The environmental effects of the spread and industrialisation of a thriving city were very great and were to remain so for well over the next hundred years. John Lee's book, published in 1933, was the last to deal with the plants of the Clyde area.

Though the first Natural History Society in Glasgow was founded as early as 1851, even as late as 1950 little thought was given to nature conservation. As a teenager trying to find the few and inadequate field guides then available, I doubt if I even knew the term. The Nature Conservancy was but one year old and naturalists' trusts, such as the Scottish Wildlife Trust, were still to come. There was only one nature reserve within the boundaries of Glasgow; this was Possil Marsh, a reserve for birds since 1930 and long famous as a noteworthy place for rare plants. Now everyone is 'green' whatever the shade. Concern for conservation locally, nationally and globally is rightly very keenly felt. Possil Marsh remains the only nature reserve inside the city. However, now there are country parks. There are nature reserves set by up by the Scottish Wildlife Trust, the Royal Society for the Protection of Birds and the Nature Conservancy Council (NCC), the Scottish part of which, at the time of writing, is called the Nature Conservancy for Scotland (NCCS). There are lesser grade reserves called Sites of Special Scientific Interest as designated by the NCCS. All are within easy reach of Glaswegians, especially the many car owners.

Local conservationists, such as members of the Glasgow Natural History Society and Glasgow Urban Wildlife Group, in and around Glasgow as much as elsewhere fight hard, at public enquiries if need be, to preserve the few remaining fragments of ancient woodland, peat-bogs and other greatly reduced habitats. In the last few years, Bishop Loch and the bogs at Commonhead and Lenzie have been saved from damaging developments by legal action. In 1989, for the first time, Glasgow District Council formally adopted a wildlife and nature conservation policy statement.

If asked to state very briefly what are my scientific interests I would respond 'Historical plant geography'. The understanding of the distribution patterns that plants show on worldwide and local scales is a fascinating topic. How are these patterns to be explained? Seed dispersal, climate, geology and soil, competition, man's activities, or some combination of those factors, as is often the case? In turning to the plants considered in this book, there are many intriguing questions. Is the tall, somewhat drab Orchid, Broad-leaved Helleborine, more common in Glasgow than any equivalent area in Britain or indeed in Europe? How did it come about that the closely related Orchid, Young's Helleborine, known only at a few places exclusively in Britain, grows on two coal bings in the Glasgow area? Why does Austrian Yellow-cress grow profusely only on rubbly waste-ground at Possilpark and only there in all of Scotland at present? Did Gipsywort colonise Glasgow only after the Forth and Clyde Canal was built? Why do all mid-Victorian cemeteries in Glasgow support Common Bistort, often in very large patches? Do soil factors alone explain why some plants, such as Primrose and Giant Bellflower, are mostly confined to the south of Glasgow, while others such as Bog-myrtle and Sheep's-bit, are entirely or mainly restricted to the north-east?

The recent objectives of the several years of field recording were several and closely linked:

1 to compile a catalogue of the native and alien wild plants of the Glasgow area;
2 to compare the plants of the built-up areas with those of the surrounding countryside and to make special study of the plants as features of the industrial and urban landscape such as bings, motorways, railways (mostly disused), canals, golf courses, gardens and cemeteries;
3 to attempt to explain the geographical patterns the plants show, and in particular, how these patterns have altered with time, especially during the last 200 years when man's activities have resulted in profound environmental changes;

4 to provide a detailed, accurate botanical basis for nature conservation within the Glasgow area.

At any time during the last 200 years, thorough botanical surveys at intervals of twenty-five years or less would have revealed even more striking changes in the plants than are obvious from studying the published works, especially the books by Thomas Hopkirk and Roger Hennedy. The 1980s survey provides much information to help understand the changes yet to come. Perhaps there should be continuous monitoring. Surveys in the 2010s and 2030s will certainly reveal many changes. If there is global warming caused by the greenhouse effect, will some of the changes be relatable to this man-made alteration of the climate of planet Earth?

Presentation and discussion of the very large amount of recently gathered information concerning hundreds of places and about 1,200 species of plants must be left to another publication. With the great benefit of the colourful illustrations provided by the skill of Elspeth Harrigan and Norman Tait, the aim of this book is to introduce some of our special urban and rural plants and their natural histories to the many members of the general public who are interested in nature conservation.

CHAPTER 2

The Glasgow Area

As defined in this book, the Glasgow area has Faifley, part of Clydebank, at the north-west corner and Gartshore, east of Kirkintilloch, at the north-east corner. Auchenraith, between Blantyre and Hamilton, is at the south-east corner and Duncarnock, south of Barrhead, is at the south-west corner. All of the city of Glasgow is included. Within the area in whole or in part are Clydebank, Bearsden, Milngavie, Bishopbriggs, Lenzie, Kirkintilloch, Muirhead, Uddingston, Bothwell, Blantyre, East Kilbride, Carmunnock, Busby, Clarkston, Williamwood, Whitecraigs, Newton Mearns, Barrhead and Renfrew. Maps 1 and 2 show the Glasgow area.

The reason why some of these places are only partly in the studied area is because of the use of the grid system on the maps published by the Ordnance Survey, as in Sheet 64 Glasgow 1:50 000 Second Series. Making up the rectangle shown in Map 1 are the whole of two ten kilometre squares (TKS), 56 and 66, and the southernmost two-fifths of TKS 57 and TKS 67 as well as the northernmost two-fifths of TKS 55 and of TKS 65. The rectangle has an area of 360 square kilometres and is divided into ninety squares, each of four square kilometres.

This Glasgow area is both low-lying and largely urban. The lie of the land and vegetation determined the choice of the northern and southern boundaries. Only two kilometres north of the north-west corner of the rectangle there is rough terrain, the Kilpatrick Hills, rising to over 300 metres above sea-level. Only a few kilometres to the south of the southern boundary, there are extensive tracts of moorland at 200 metres and more above sea-level. The western and eastern boundaries were chosen to make the project feasible in the time and with the limited number of skilled helpers available. These boundaries could have been placed further west to have included all of Paisley and further east to have included Airdrie, Wishaw and Motherwell and the low-lying, urban character of the area would have remained.

There are ten political districts within the rectangle. By far the biggest is the City of Glasgow District which covers more than half of the area, about 200 of the 360 square kilometres. The other Districts in descending order of area are Strathkelvin (about sixty-three square kilometers), East Kilbride, Eastwood, Renfrew (each about twenty square kilometers), Bearsden and Milngavie (about fifteen square kilometres), Hamilton (about thirteen square kilometres), Clydebank (about six square kilometres), Motherwell (about eight square kilometres) and Monklands (less than one square kilometre).

The highest points are all in the far south. Only a few metres outside the south-west corner, Duncarnock reaches 204 metres above sea-level. The Cathkin Braes do not quite reach 200 metres. Nearby is the highest point in the area at South Cathkin Farm: 213 metres. To the east, Dechmont Hill, the prominent hill nearest the south-east corner, reaches only 183 metres.

Near the north-west corner, Windyhill Golf Course rises to over 160 metres and in the north-east at Gartshore, the ground extends to little over 100 metres. About ninety per cent of the area lies below 100 metres above sea-level and about thirty-five per cent lies below 50 metres.

Almost all the bedrock of the Glasgow area was formed during Carboniferous times 300 and more million years ago. Over most of the area the rocks are sedimentary, laid down under water: coal measures, millstone grit and limestones, though the content of lime is not high. The only extensive outcrops of volcanic rocks are in the south; at Duncarnock and the Cathkin Braes the rocks are basaltic.

The ice sheets of the Ice Ages moulded the landscape and covered much of the solid rock with mixtures of clay, silt, sand, gravel and boulders, some of which came from far away from Glasgow. The glaciers dumped the very numerous, small, smooth hills which are a prominent feature of the city and its surroundings. These hills, called *drumlins*, were formed somewhat less than 30,000 years ago, when the glaciers of the last Ice Age were very large. The Necropolis hill is not a drumlin but a *craig-and-tail*, another landform produced by the last Ice Age, as discussed by my geological colleague Graham Jardine. Some of the drumlins protrude through the sediments settled out when the large basin in which Glasgow sits was twice flooded by the sea, first about 13,000 years ago and second about 8,000 years ago. These inundations produced the flat terrain that rises little above sea-level and widens irregularly from east to west within the Glasgow area.

The earliest men in the area were Stone Age hunters and gatherers, present in very small numbers. When they arrived several thousand years ago, deciduous woodland had already grown over the dry ground in the area for thousands of years since the end of the last Ice Age about 10,000 years ago. By the time of building the cathedral, some 800 years ago, the woodlands were very much reduced by human activities. At that time, Glasgow was very small and not until the late eighteenth century and into the nineteenth century did the great expansion of the city take place. The story of the development of Glasgow is well and succinctly told by Andrew Gibb and in lighter terms by Charles Oakley.

Fifty-six of the ninety squares in the Glasgow area are *urban*. A square defined as urban is between half and entirely built up by housing, industry, rubbly wasteground, railways (both active and disused), roads including motorways, cemeteries and bings. The thirty-four *rural* squares are less than half so built up; they are shown white and the urban squares stippled on the Maps 1 and 3 to 10.

PLATE 1

The canal towpath at Possil in mid August

× 0.4 natural size

All these plants can be seen without leaving the towpath. A Victorian out for a stroll would probably not have seen Rosebay Willowherb (pink flowers), a rare plant 100 years ago. In size, shape and colour, both the leaves and the flowers resemble those of the shrub Nerium oleander, *Rosebay, common in Mediterranean countries; hence the name. Great Reedmace (centre) has spread a lot around Glasgow in the last 150 years. Like many water and marsh plants, Great Reedmace has a very wide spread in the world. The fruiting heads are often gathered as long-lasting decorations. If you have the patience it would be more fun to gather the pollen grains to make into a gingerbread-like food. First find a Californian*

Redskin or a New Zealand Maori who may tell you how to do it.

Gypsywort (bottom left) grows throughout the canal. It is a very old story that gypsies stained their skins dark with this plant. As Geoffrey Grigson stated 'Whether any English gypsy [or Scottish tinker one might add] has ever been known to stain himself with Lycopus europaeus *is another matter. It does give a good black stain.' The last remark has been experimentally proved by Su Grierson who also obtained 'chocolate' and 'lettuce green' colours.*

The yellow flowers are those of a Hawkweed (Hieracium)*, belonging to a group of plants that few, even among botanists, can identify correctly. What hawks have to do with Hawkweeds has been explained by Humphrey Gilbert-Carter. An ancient name,* Hieracium *comes from the Greek, referring to hawks. He stated unexplicitly: 'this bird was said to use them for diseases of the eye.' The original story comes from the credulous, prolific writer, the Roman Pliny the Elder who wrote 'hawks tear it apart and wet their eyes with the juice, so dispelling dimness of sight, when it comes to them.'*

Mints are another difficult group. There are at least eleven sorts, four species and seven hybrids. Some of the hybrids are often grown and being too rampageous are tossed out by gardeners. To the right of centre is a spontaneous hybrid, Whorled Mint, a cross between Corn and Water Mints.

Despite its name, Common Reed (right) is not very common around Glasgow. It is the tallest non-woody native British plant, often reaching two metres high and sometimes more. In more congenial climates, as in the Danube delta and in Malta, its height can be much greater.

PLATE 1

PLATE 2

PLATE 3

PLATE 4

PLATE 2

Prickly Heath near the city centre

× *0.7 natural size*

PLATE 3

Common Spotted Orchid near the city centre

× *0.6 natural size*

PLATE 4

Northern Marsh Orchid near the city centre

× *0.6 natural size*

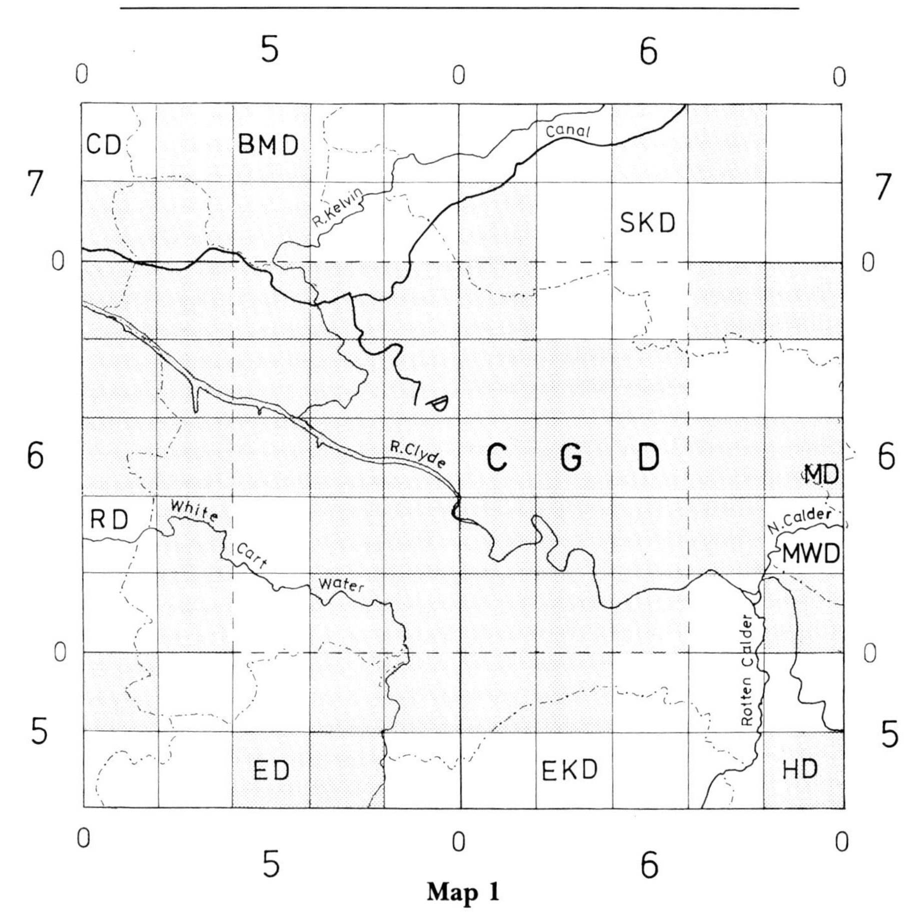

Map 1

The Glasgow area. The dot-dash lines show the district boundaries. CGD = City of Glasgow District BMD = Bearsden and Milngavie District CD = Clydebank District ED = Eastwood District EKD = East Kilbride District HD = Hamilton District MD = Monklands District MWD = Motherwell District. The dashed lines separate the 30 central squares from the 60 marginal ones. The River Clyde and its tributaries are

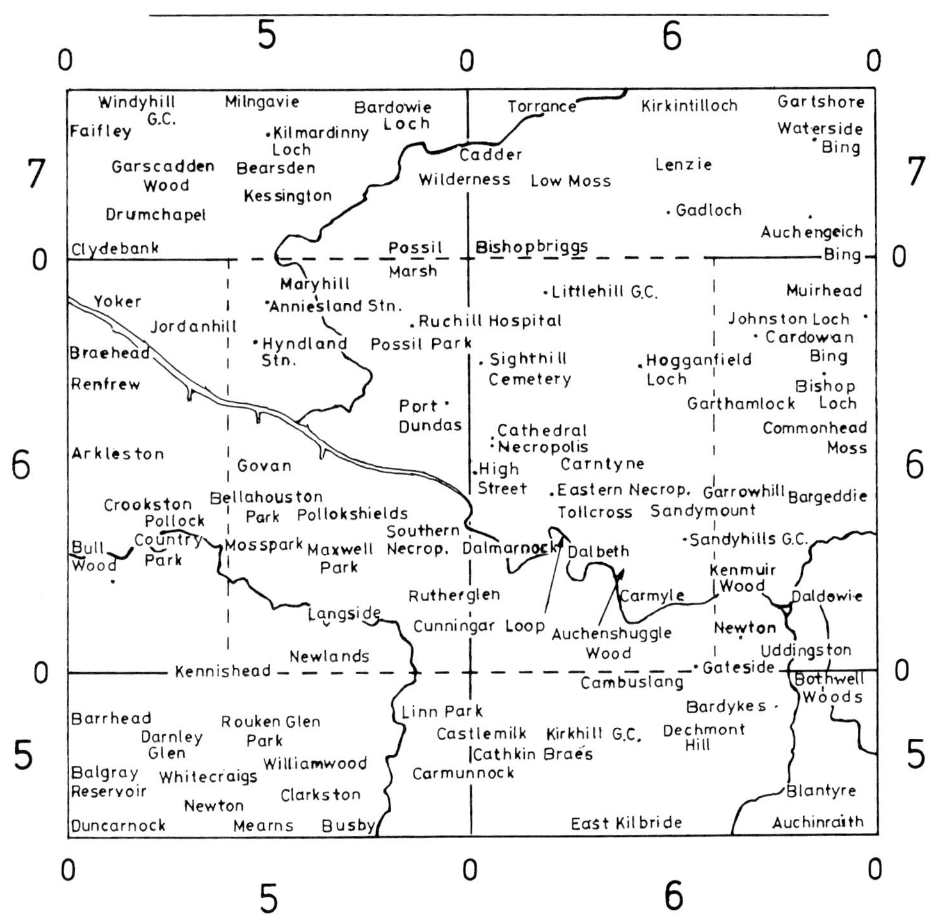

shown as is the Forth and Clyde Canal with the Glasgow Branch reaching to Port Dundas. The squares stippled dark are urban and those left white are rural.

Map 2

Some of the places mentioned in the text.

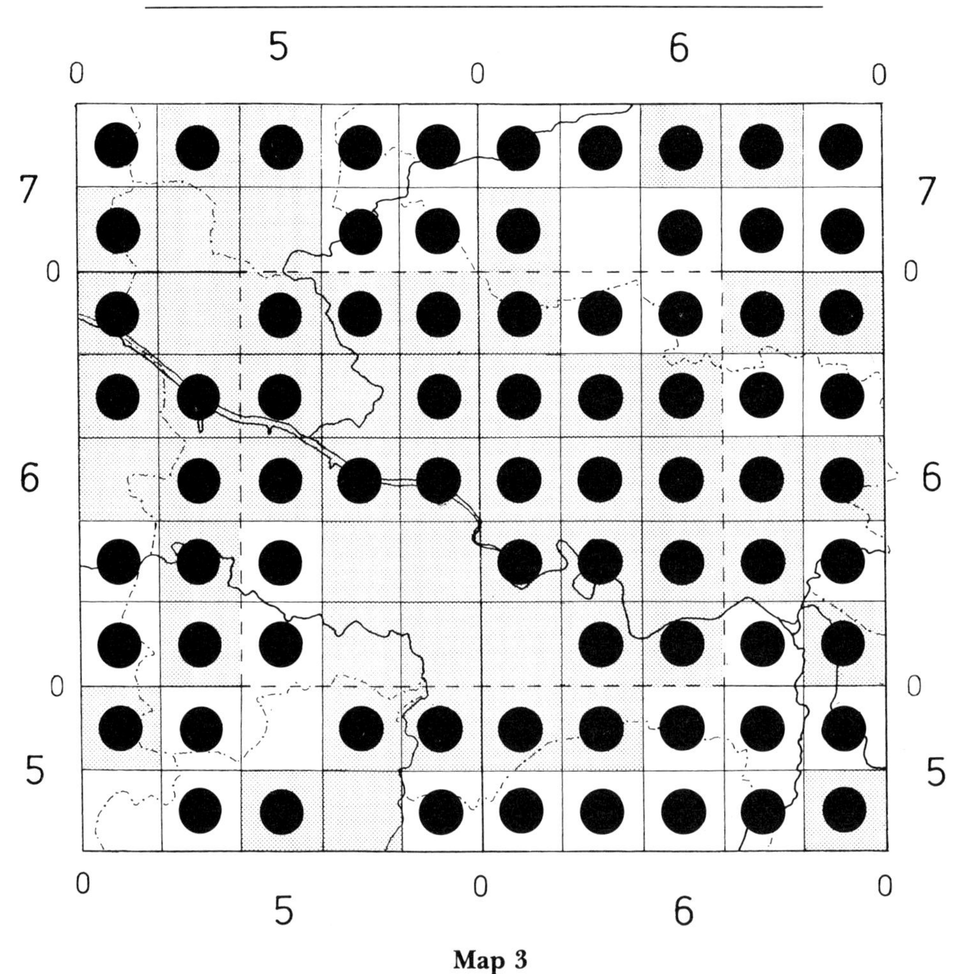

Map 3

An easy-goer: Common Spotted Orchid (Dactylorhiza fuchsii).
The blackened circles indicate presence but do not indicate abundance; only one to very many plants may have been recorded in a square.

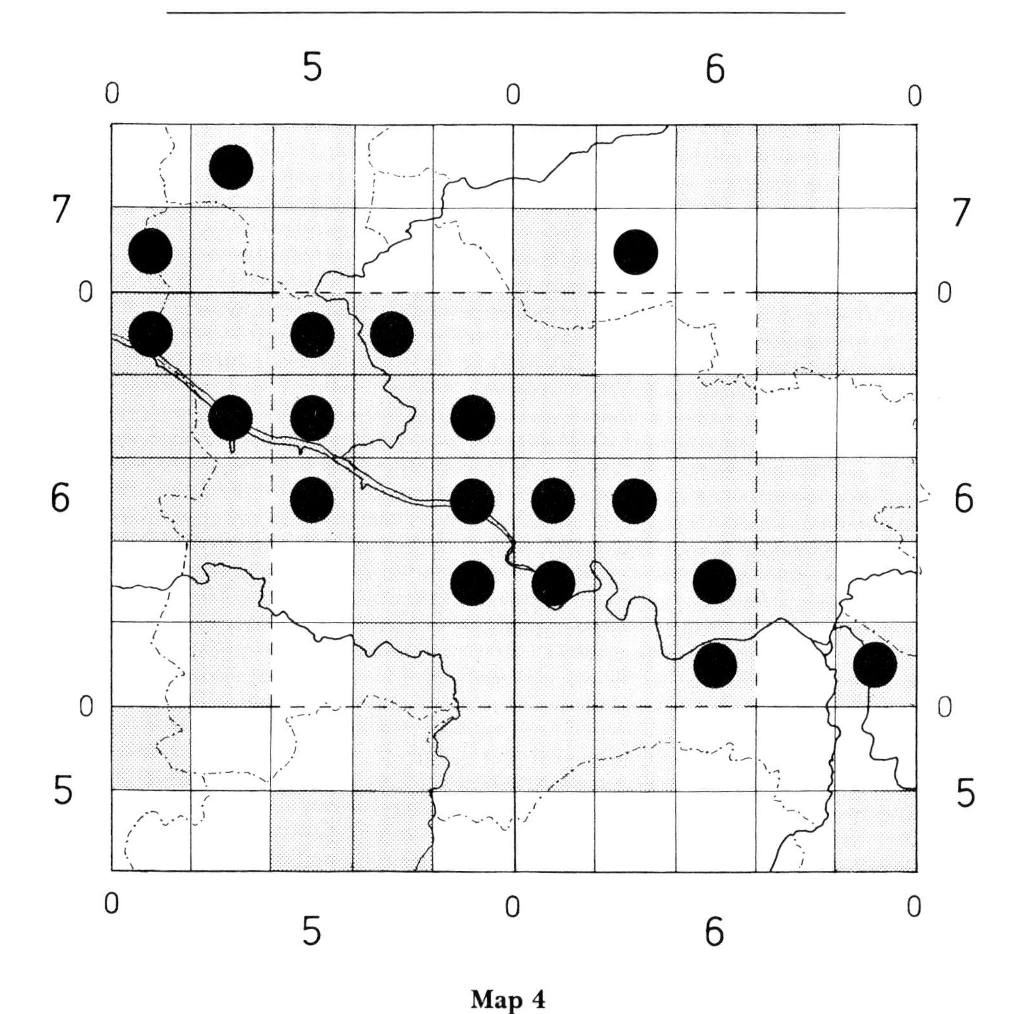

Map 4

The city dweller Butterfly-bush (Buddleja davidii) *with a diagonal pattern mostly at the lowest altitudes. The habitat of the sole rural square is the abandoned railway sidings at Cadder Yard.*

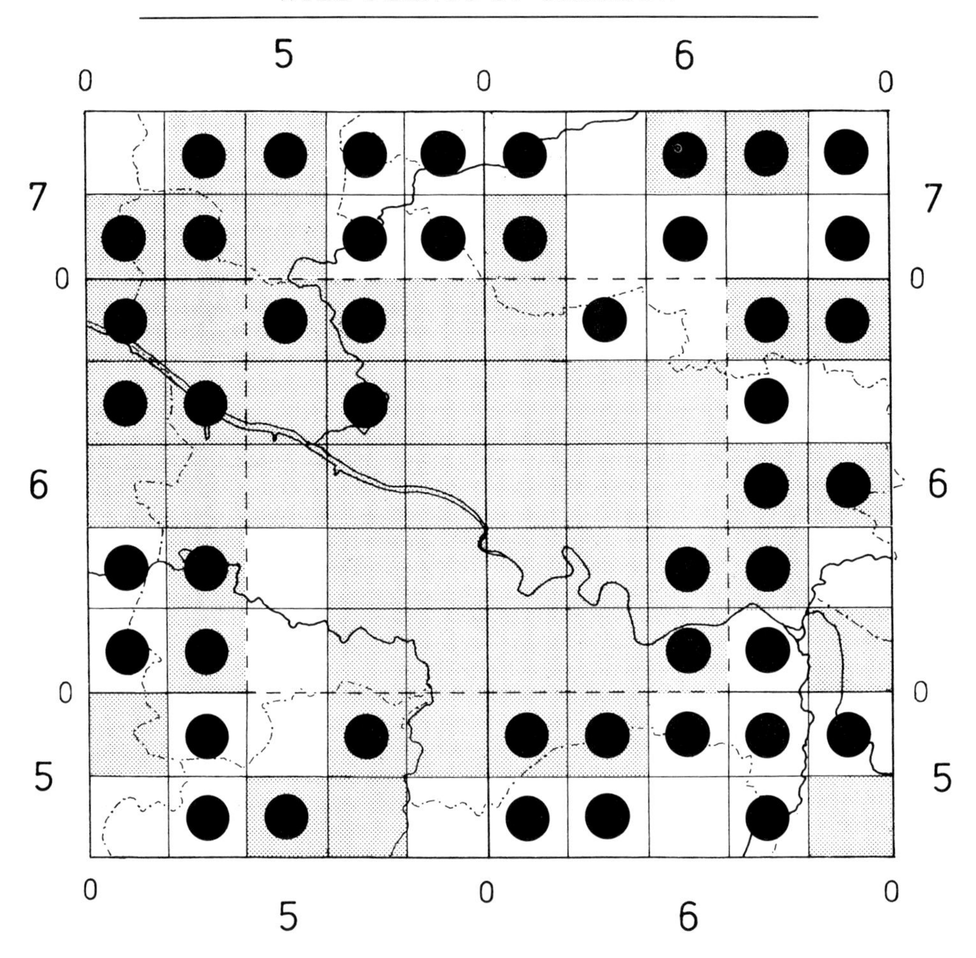

Map 5

The country cousin *Mouse-ear Hawkweed* (Hieracium pilosella).
There is a large central gap.

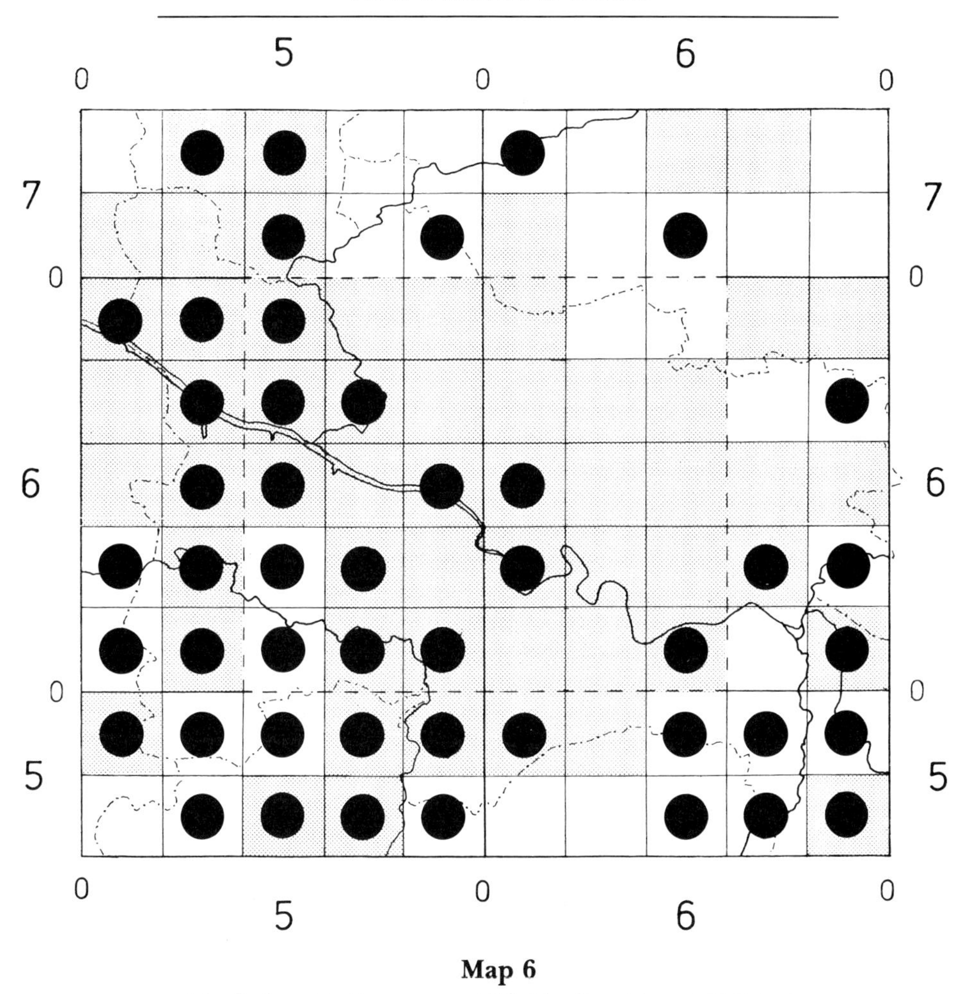

Map 6

The L-shaped pattern as exemplified by Broad-leaved Helleborine (Epipactis helleborine). *If the pattern were plotted on a finer scale, say 1km squares, the scarcity in the north-east and the commonness of this Orchid in the south and west would be even clearer.*

PLATE 5

Glasgow streets in mid September

× *0.4 natural size*

In recent years there has been a great deal of amenity planting in pedestrian precincts, housing developments and along the walkways by the Clyde, Kelvin and Cart. Most of these alien plants never spread away from these sites and some can never do so. The Limes are a good example. Our July is very rarely, if ever, warm enough for the flowers to be fertilised and set good seed. Some others set abundant ripe seed and spread.

Rowan and its relatives are among the most frequently planted small trees of the Glasgow area. Because some regenerate freely in gardens, parks and wasteground they have become an established feature of the city's wild flora.

Rowan can cross with the Whitebeams. Centre-left is the hybrid of Rowan and Common Whitebeam (Sorbus x thuringiaca). *Occasionally planted as in Victoria Park, Hyndland Bowling Club and Milngavie pedestrian precinct, it has only one or two pairs of leaflets at the base of each leaf, then some deep lobes and a tapering more or less deeply toothed apex. Self-sown seedlings have not been found around Glasgow or indeed elsewhere. In the centre is another hybrid, a cross between Rowan and Swedish Whitebeam* (Sorbus x pinnatifida). *In the Glasgow area only one obviously self-sown tree was known; it grew from the base of a wall in Novar Lane, Hyndland where it was discovered by June Mackay. It may have been the only such wild hybrid in Scotland and there seem to be only few reports from England. Nearby grew the putative parent trees as well as Common Whitebeam, all regenerating. The leaves have two or three basal leaflets, then a lobed, short, somewhat triangular apex.*

PLATE 5

PLATE 6

PLATE 6
Pineappleweed
× *0.9 natural size*

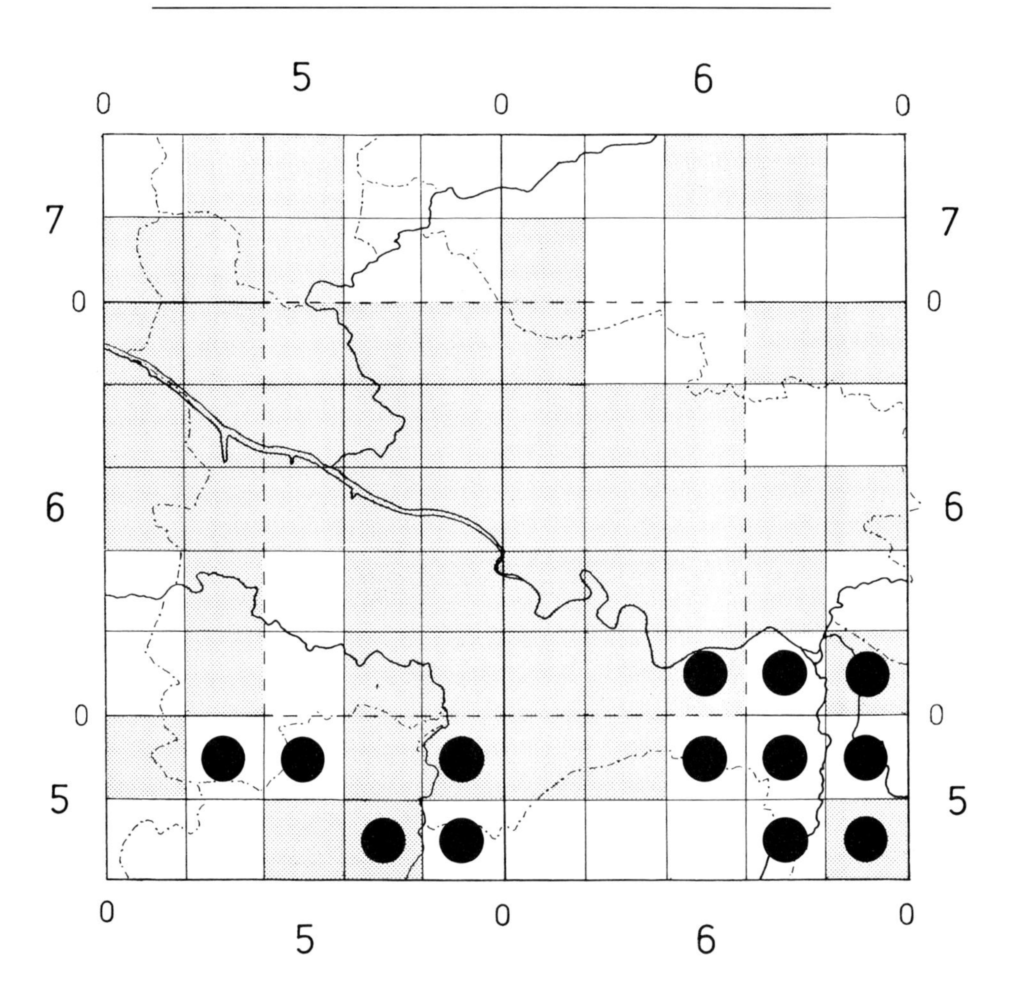

Map 7

The southern pattern as exemplified by Hard Shieldfern
(Polystichum aculeatum).

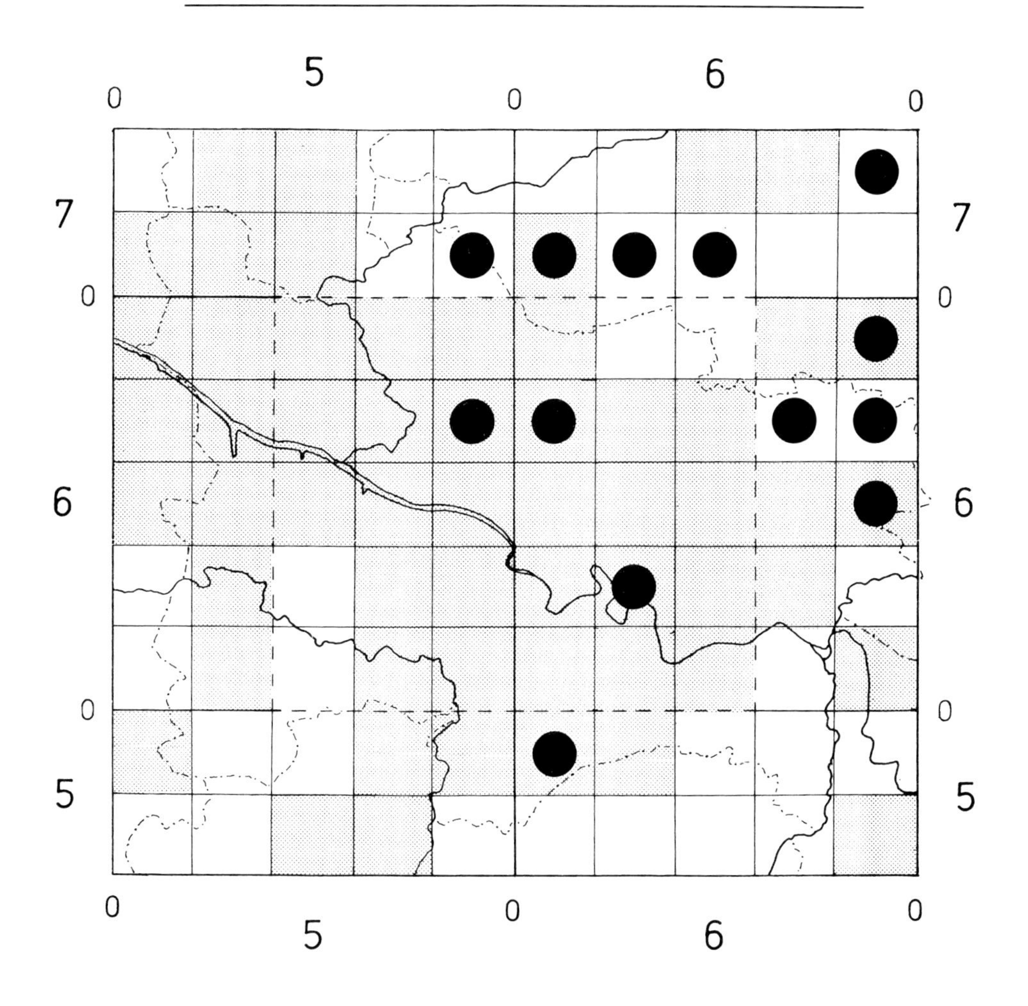

Map 8
The north-eastern pattern shown by Hare's-tail Cottongrass
(Eriophorum vaginatum).

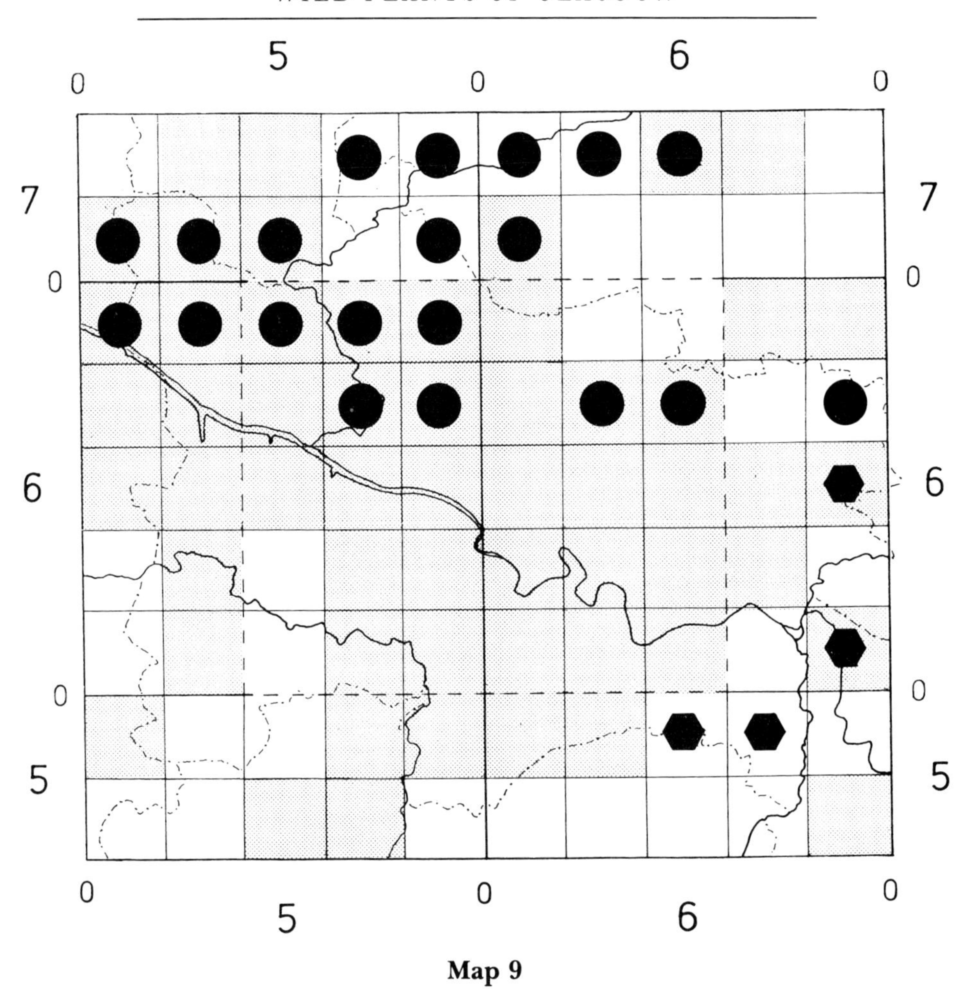

Map 9

The circles show Gipsywort (Lycopus europaeus) *with the linear patterns following the canals and the hexagons show the south-eastern pattern of Rough Chervil* (Chaerophyllum temulentum).

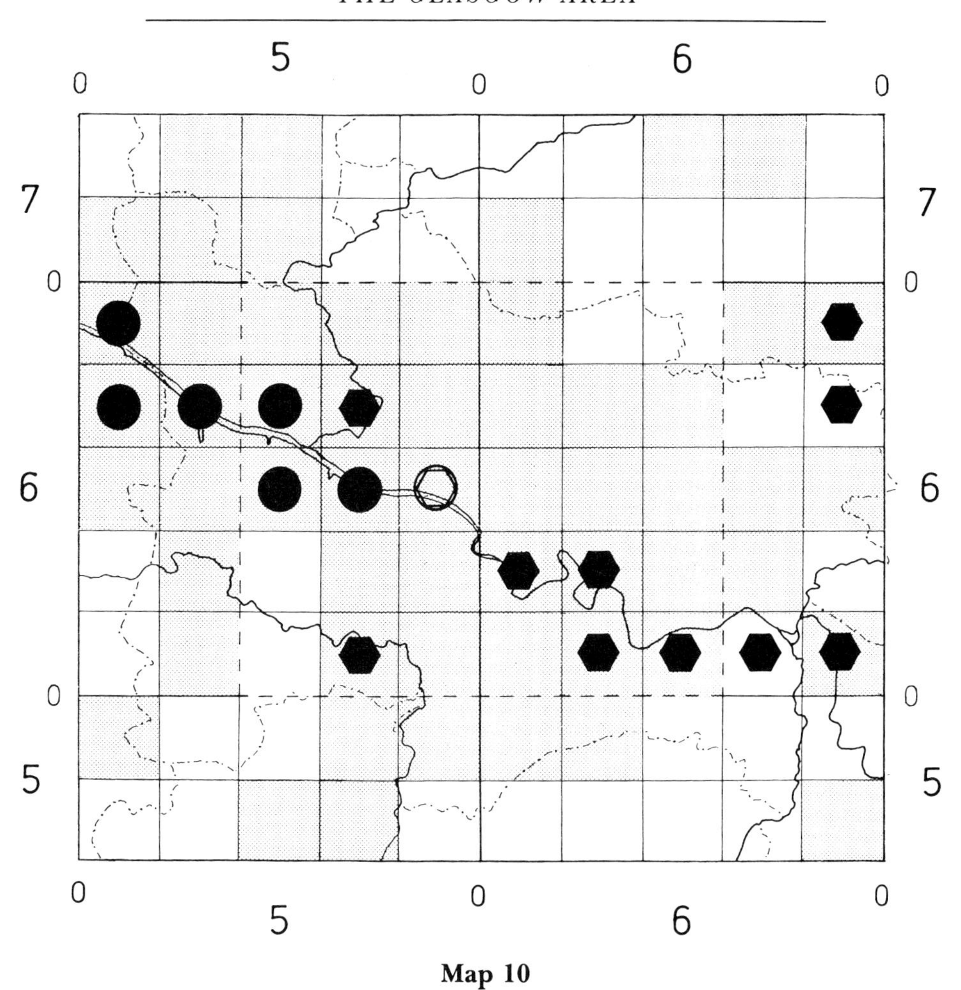

Map 10

The circles show Garden Angelica (Angelica archaneglica) *which grows exclusively along the Clyde and the hexagons show Fennel Pondweed* (Potamogeton pectinatus) *which grows largely but not entirely along the Clyde.*

CHAPTER 3

Surveys, Old and New

By far the earliest list of plants relevant to the Glasgow area was made some time between 1549 and 1592. It was written on the back flyleaf of a copy of Fuchs' *Herbal* (a book dealing with the medicinal properties of plants). The writer was Mark Jameson, a clergyman at Glasgow Cathedral, physician and deputy rector of the university. Under the heading, 'To be sett or sown in ye garding', Jameson wrote the names of twenty-two plants, some of which, such as Yarrow, Common Valerian, Black Spleenwort, Carrot, Parsnip, Hart's-tongue and Pot Marigold, might have escaped from cultivation at the cathedral, the likely location of Jameson's garden of medicinal plants. In the *Scottish Medical Journal* a few years ago Bill Gauld and I discussed Jameson's list in detail.

One of the most famous English naturalists, John Ray, visited Glasgow in 1661 and admired the university but, sadly made no botanical records. From the second half of the eighteenth century come the first substantial plant lists. John Hope (1725–86, professor at Edinburgh) stimulated his students and collaborators to study the Scottish flora. He had an herbarium (collection of pressed and dried plants) containing perhaps as many as fourteen specimens that had been gathered from the Glasgow area. These are the first relevant records made in a modern way.

A Walking Laboratory: Rev David Ure (1749–98)

> Whether travelling to gratify his own curiosity, or to execute any commission, it was always on foot. Tho' short of stature, he enjoyed a sound constitution, and a vigorous structure of body. He often carried bread and cheese in his pocket, and enjoyed his repast beside the cooling spring. . . . When his circumstances could afford it, he would repair to the village ale-house, and enjoy his favourite luxury, a glass of ale. His great-coat was furnished with a large pocket, in which he stowed such minerals, or other curiosities, as had attracted his notice. . . . He carried a tin box for stowing curious plants; a large cudgel, armed with steel, so as to serve both as a spade and pick-axe; a few small chisels, and other tools; a blow pipe, with its appurtenances; a small liquid chemical apparatus; optical instruments, &c. &c. so that his friends used to call him a walking shop, or laboratory. In this way he braved all weathers, and heat or cold, wet or dry, seemed equally indifferent to him. He was a patient observer and accurate describer of nature. His descriptions were always taken down on the spot, in a hieroglyphical species of shorthand invented by himself, and which, it is to be regretted, no one but himself understood.

That there is no likeness of this lad o'pairts is a great pity, as Rev Headrick's description of him, ready for field work, makes clear. David Ure, son of a Glasgow weaver, and graduate of Glasgow University, was very able, energetic, and good-humoured. He was the first naturalist to publish a list of plants from the Clyde area. The list is printed in his wide-ranging book *The History of Rutherglen and East Kilbride* which appeared in 1793. Full of interest, it is a fine book containing, among other illustrations, drawings of Carboniferous fossils of animals and plants.

Under the heading 'List of Scarce Indigenous Plants in Rutherglen and Kilbride' there are 151 plant names with localities. The vast majority are Flowering Plants and Ferns and at least ninety David Ure had seen at places in the south-east of the Glasgow area or nearby to the south.

Pioneer Botanist of Clydesdale: Thomas Hopkirk (1785–1841)

Born of a well-to-do family, Thomas Hopkirk was the cultivator of nearly 3,400 plants at Dalbeth and was instrumental in the founding of the Botanic Gardens. He entered Glasgow University in 1800 but never graduated. However, in 1835, the university conferred on him an honorary LL.D in recognition of his scientific achievements. Apart from his enthusiasm for gardening, he was a devoted field botanist at a time when man-traps and spring guns were a hazard to anyone straying into game preserves. When only twenty-eight years old, he published in 1813 *Flora Glottiana. A Catalogue of the Indigenous Plants on the Banks of the Clyde and in the Neighbourhood of the City of Glasgow*. His critical approach shines through this small book, one of the earliest local Floras (a Flora is a book about the plants of a particular area).

In explaining his area of study, Hopkirk stated '. . . I have in general confined myself to the country within a few miles on

both sides of the River Clyde, from its Falls, above Lanark, to its junction with the sea some miles below the Town of Dumbarton.' *Flora Glottiana* contains hundreds of first records of plants of west-central Scotland, very many relevant to the Glasgow area. It gives the localities of many plants now rare, declining or extinct.

As is the case with David Ure, it is not known if Thomas Hopkirk made an herbarium. Consequently, there is no way of checking their records, though only very few are questionable.

Published in 1821, *Flora Scotica* by Regius Professor William Jackson Hooker contains only a very small number of first records additional to those of Thomas Hopkirk and relevant to the Glasgow area. The same comment is true of both *Indigenous Plants of Lanarkshire* by Rev Patrick and published in 1831 and the botanical lists from excursions made by Professor James Hutton Balfour.

Failed Businessman, Successful Author:
Roger Hennedy (1809–76)

With a claim to being one of the most successful local Floras ever written, Roger Hennedy's *The Clydesdale Flora. A Description of the Flowering Plants and Ferns of the Clyde District* ran to no less than five editions, two of them published after the author's death. Hennedy became professor at the Andersonian University (now the University of Strathclyde) only in 1863 after many years in the textile business when his 'devotion to flowers occupied every spare moment of his time'. He was a very conscientious field worker and teacher, sometimes unpaid, before attaining his academic post.

Another once well-known author, whose little book *Rambles Round Glasgow* was very popular, was a friend of Roger

Hennedy. Describing a ramble together near Provanmill, Hugh Macdonald wrote 'Our flower-loving friend is now in all his glory, poking and prying along the vegetable fringe that skirts the path. Every now and then we are startled by his exclamations of delight, as some specimen of more than ordinary beauty meets his gaze.' On another occasion, a collier, seeing Roger Hennedy's activities exclaimed, 'That wee ane's daft, he's clean gyte; see he's gathering glaur and pittin'd in a bottle.' This was a sample of mud for the study of microscopic algae, another of Roger Hennedy's passions.

Unlike David Ure, Roger Hennedy never entered a public house, at any rate not with his students. He never allowed himself to be photographed nor to have a portrait painted. The sketch reproduced here was made for the *In Memoriam* edition of the *Flora* published in 1878.

Roger Hennedy's herbarium is now housed in the natural history section of Kelvingrove Museum.

Another Straight-laced Victorian: John Ramsay Lee (1868–1959)

With an interest in field botany extending back to his teens, John Lee was well into his sixties when he published *The Flora of the Clyde Area* in 1933. The very extensive region that is the whole drainage basin of the Clyde, including the Firth, is covered. Unadorned by as much as a map, it is a drab book which does no justice to the lifelong enthusiasm of its author.

A cashier in a city warehouse, devoted churchman, and bachelor long resident in Dennistoun, John Lee was greatly respected by all his contemporaries, including the author. We were once in the field together at Lochwinnoch in 1955; I was

eighteen and he was eighty-seven, then and for many years previously a benign figure who held my rapt attention by recalling the days in the late nineteenth century when Rosebay Willowherb was a rare plant.

John Lee made a large herbarium of fine specimens, though lacking all habitat information. His material has been incorporated into the British herbarium in the Botany Department of Glasgow University.

Apart from these few authors of books, from the mid nineteenth century right up to the present day many other naturalists have studied the plants of the Glasgow area and west central Scotland. Their findings have been published in periodicals such as *The Glasgow Naturalist*, the journal of the Glasgow Natural History Society.

Field Work in the 1980s

The *Atlas of the British Flora* was a landmark in the many achievements of the long tradition of field botany in the British Isles. In the mid and late 1950s the Botanical Society of the British Isles backed a project to record British plants on the grid system of the Ordnance Survey. The basic recording unit was the ten-kilometre square. Master-minded by professionals, the project depended on the skill and enthusiasm of hundreds of amateur field botanists, probably more numerous in Britain than anywhere. The culmination in 1962 was the *Atlas*, edited by Frank Perring and Max Walters. In this large volume there are many hundreds of dot maps revealing the many distribution patterns shown by plants on a British Isles scale.

A major advantage of working by the grid square method is that, to be thorough, the recorder is forced to go to all areas, no matter how uninviting a particular square may appear to be. Often the appearance is deceptive and there are unanticipated rewards. To carry out a study of the Glasgow area, the ten-kilometre scale was too coarse to be worthwhile. The finer the scale the more is revealed. One-kilometre squares would yield much more significant detail than ten-kilometre squares. However, there are 360 one-kilometre squares, too many to be tackled. The ninety squares, each of four square kilometres, were chosen as a feasible number which would give a great deal of useful information.

From the beginning of plant recording in the Glasgow area to the present about 1,200 species have been found. Any one of these 1,200 wild plants can be *native* or *alien* or both. A plant native in the studied area has arrived by natural means of dispersal, that is without being transported by man. An alien plant has been brought to the area by man, either intentionally or unintentionally. Cultivated plants were recorded only if they were seen to be spreading well away from their original positions of growth.

The 1980s survey was carried out by a diversity of recorders: many were members of the Glasgow Natural History Society, some were members of Adult Education classes taught by the author and some were botany students at Glasgow University. Others were graduate employees of a Manpower Services Commission Community Programme, *Botanical Surveys of Glasgow*, lasting two years from 1985 to 1987; the work supervised by Henry Noltie and Keith Watson, both very knowledgeable field botanists, concentrated on disused railways, motorways, coal bings, golf courses and lochs.

CHAPTER 4

Patterns of Plants around Glasgow

No two plant species can occupy exactly the same space and so logically there are as many geographical patterns as there are species. It is impossible to make maps at a fine enough scale to reveal such minutely unique patterns, except for extremely small areas. At the scales convenient to use for local regions, countries or the planet, broadly similar patterns shared by species recur over and over again and need explanation. This is a task which has happily engaged plant geographers for many years.

The ten maps on pages 30–35 and 40–43 have been chosen to show something of the diversity of geographical patterns that plants show around Glasgow. The map of Common Spotted Orchid is typical of approximately 100 species that have been recorded in seventy-five or more of the ninety squares. Some of the most familiar plants occur in all ninety squares. Examples are Annual Meadow-grass, Colt's-foot, Creeping Buttercup, Daisy, Field Horsetail, Ground-elder, Groundsel and Rosebay Willow-herb (see Tables 1 and 2). These can be thought of as *easy-goers*.

The ninety squares can be divided up into sixty marginal squares (two deep all round) and thirty central ones. All but four of the central squares are urban whereas twenty-nine of the sixty

PLATE 7
Dame's-violet
× *0.6 natural size*

PLATE 7

Figure 1 *Giant Hogweed along the Kelvin at Partick Bridge.*

Figure 2 *Garden Angelica at Braehead, Renfrew.*

Figure 3 *Hybrid Yellow Waterlily (Photograph J H Dickson).*

Figure 4 *Mudwort at Balgray Reservoir.*

Figure 5 *Few-flowered Leek at Kelvinbridge.*

Figure 6 *The Clyde at Dalmarnock with Indian Balsam and Black Mustard. The wild Fig grows in the background.*

PLATE 8

PLATE 8

Slag bings and wasteground at Hallside Steelworks in early August

$\times$ *0.4 natural size*

The striking blue-purple flowers are those of Viper's Bugloss, a plant that can reach one metre tall. Bugloss comes from the Greek for 'ox-tongued'. The connection with snakes goes back at least to the book on medicinal plants written in the second century AD by the Greek doctor Dioscorides. The first English translation of this very influential work was not until 1655 by John Goodyear who wrote: 'Ye flowers by the leaves of a purple colour, in which also is ye seed like ye head of a viper: ye root thinner than a finger, somewhat black which not only helps those already bitten by serpents being drank with wine, but also makes those which drink it before, unbitten.'

To the left is a stem of the yellow-flowered, felty-leaved Great Mullein, a plant which can reach two metres tall. Geoffrey Grigson in his book A Dictionary of Plant Names *explained that Mullein comes originally from the Latin* mollis *meaning soft, a reference to the texture of the leaves.*

Middle right are two stems of Weld, a plant which can attain one metre. Weld is a Germanic name. At the back are the dark flowers of Woody Nightshade, a plant which can straggle or climb several metres. Nightshade probably alludes to the harmful berries and the often shady habitats of Deadly Nightshade. To the bottom right is the yellow-flowered Large-toothed Hawkweed.

Table 1. Flora of Hyndland Stations (old and new)

A	Alsike Clover	FD	Henbane
UA	American Willowherb		Hop Trefoil
U	Annual Meadow-grass	F	Hybrid Rowan
U	Ash	RFD	Isle of Man Cabbage
U	Bird's-foot Trefoil	UA	Japanese Knotweed
R	Bladder Campion	U	Knotgrass
U	Bramble	FD	Large Hop-trefoil
UW	Broad Buckler-fern		Maidenhair Spleenwort
U	Broad-leaved Docken	UW	Male-fern
A	Butterfly-bush		Marsh Cudweed
A	Canadian Goldenrod		Marsh Yellowcress
U	Cock's-foot	UA	Michaelmas Daisy
U	Common Bent	U	Mugwort
U	Common Couch	A	Opium Poppy
U	Common Knapweed	A	Oxford Ragwort
U	Common Mouse-ear		Pale Toadflax
U	Common Nettle		Perforate St John's-wort
U	Common Orache		Raspberry
R	Common Toadflax	AW	Red-berried Elder
	Common Whitebeam	U	Red Clover
U	Cow Parsley	U	Red Fescue
U	Creeping Buttercup	U	Reed Canary-grass
UR	Creeping Cinquefoil	U	Ribwort Plantain
U	Curled Docken	U	Rosebay Willowherb
U	Dandelion	FDA	Rough Bent
U	Elder	U	Rowan
U	False Oat-grass	U	Scentless Mayweed
	Fat Hen	U	Sheep's Sorrel
R	Field Bindweed	U	Shepherd's Purse
U	Field Horsetail	U	Silver Birch
U	Field Thistle	U	Silverweed
W	Foxglove	R	Small Toadflax
A	Garden Lupin	U	Smooth Sow-thistle
U	Goat Willow	U	Spear Thistle
U	Great Willowherb	R	Sticky Grounded
U	Greater Willowherb	FD	Swamp Meadow-grass
U	Groundsel	A	Swedish Whitebeam
FD	Hare's-foot Clover	UA	Sycamore

Table 1. Continued

	Tansy	W	Wall-rue
FDA	Ternate-leaved Cinquefoil		Welted Thistle
	Tormentil		White Clover
D	Treacle Mustard		Woody Nightshade
U	Tufted Hair-grass		Yarrow
U	Tufted Vetch	U	Yorkshire Fog
A	Wall Cotoneaster		

U = present in 75 to 90 of the Glasgow area squares. R = strongly connected with railways. F = in only one or a few squares. D = Disappeared by 1990 or before. A = alien in Britain. W = only on or at the foot of walls.

Table 2. Flora of Rubbly Wasteground at High Street

Annuals (or biennials) 43 species

UP	Annual Meadow-grass	A	Oxford Ragwort
	Annual Pearlwort	UA	Pineappleweed
	Barren Brome	U	Prickly Sow-thistle
A	Bread Wheat		Radiate Groundsel
U	Charlock		Rat's-tail Fescue
UP	Chickweed	U	Redshank
	Common Field Speedwell	U	Scentless Mayweed
	Equal-leaved Knotgrass	U	Shepherd's Purse
	Eyebright		Shining Crane's-bill
	Fairy Flax		Small Toadflax
UP	Field Forget-me-not	U	Smooth Sow-thistle
	Great Mullein	W	Snapdragon
	Groundsel	U	Spear Thistle
	Hairy Bitter-cress		Sticky Groundsel
	Hemlock	U	Sticky Willie
	Herb Robert	PA	Tall Rocket
U	Hogweed		Thalecress
	Lesser Burdock		Wall Speedwell
	Lesser Trefoil		Wavy Bitter-cress
	Long-headed Poppy		Weld
	Marsh Cudweed		Welted Thistle
A	Oil-seed Rape		

Perennials (non-woody) 59 species

A	Alsike Clover	UP	Meadow Buttercup
PA	American Willowherb	UP	Meadow Thistle
U	Bird's-foot Trefoil	UA	Michaelmas Daisy
UW	Broad Buckler-fern	U	Mugwort
W	Bracken	U	Oxeye Daisy
U	Broad-leaved Docken	U	Perennial Ryegrass
U	Cat's-ear	U	Perennial Sow-thistle
U	Cock's-foot		Perforate St John's-wort
U	Colt's-foot	A	Potato
UP	Common Bent	U	Procumbent Pearlwort
U	Common Couch	U	Red Clover
U	Common Figwort	U	Ribwort Plantain
U	Common Hawkweed	UP	Rosebay Willowherb
U	Common Mouse-ear	PA	Rough Bent
U	Common Nettle		Sand Sedge
U	Common Ragwort	U	Self-heal
	Common Toadflax		Sheep's Sorrel
U	Cow Parsley	U	Silverweed
U	Creeping Bent	A	Slender Rush
U	Creeping Buttercup	A	Slender Speedwell
U	Creeping Thistle	U	Smooth Meadow-grass
U	Curled Dock		Tansy
U	Dandelion	U	Tufted Hair-grass
	Feverfew	U	Tufted Vetch
U	Field Horsetail		White Campion
U	Greater Plantain	U	White Clover
W	Hart's-tongue		Wild Strawberry
UPA	Japanese Knotweed	U	Yarrow
UW	Lady-fern	U	Yorkshire Fog
U	Male-fern		

Shrubs and Trees 13 species

U	Bramble	U	Raspberry
	Broom		Rowan
PA	Butterfly-bush		Silver Birch
	Common Whitebeam	A	Swedish Whitebeam
	Downy Birch	UA	Sycamore
U	Elder	U	Wych Elm
UP	Goat Willow		

U = present in 75 to 90 Glasgow area squares. P = in large patches or as very numerous individuals. A = alien in Britain. W = only on or at the foot of walls.

marginal ones are rural (Map 1). The pattern shown by Butterfly-bush (Map 4) is an outstanding example of a predominantly central one that is almost entirely urban. Other plants with strongly central, urban distributions are Barren Brome, Giant Hogweed, Great Mullein, Indian Balsam, Hungarian Brome, Oxford Ragwort, Ribbed Melilot, Rough Bent, Tall Rocket, Teasel and Wild Parsnip. These can be thought of as *city dwellers*.

The *country cousins* live entirely or mainly round the sixty marginal squares. Good examples are Mouse-ear Hawkweed (Map 5), Bog Stitchwort, Cuckooflower, Devil's-bit Scabious, Floating Reed-grass, Giant Fescue, Marsh Bedstraw, Marsh Marigold, Ramsons, Red Campion and Wood Avens.

A north-west to south-west to south-east pattern is shown by the Orchid, Broad-leaved Helleborine (Map 6). Other plants showing this roughly L-shaped pattern with scarcity or absence in the north-east are False Brome, Hazel, Wood Docken and Wood Pimpernel. Some plants have a mainly or entirely southern distribution. Good examples are Hard Shield-fern (Map 7), Bearded Couch, Crosswort, Mountain Pansy, Primrose, Pyrenean Valerian and Sanicle.

Another pattern is the north-eastern one, which is the converse of the L-shaped one. Hare's-tail Cottongrass (Map 8) is a good example, as are Bog-myrtle, Round-leaved Sundew, Cranberry and Cross-leaved Heath.

A small group display an exclusively or mainly south-eastern pattern. Rough Chervil (Map 9), White Deadnettle and Viper's Bugloss are examples.

Narrow patterns in lines across the middle and north of the area are shown by Garden Angelica (Map 10), Bennett's and Fennel Pondweeds (Map 10), Gypsywort (Map 9), Lesser

Water-parsnip, Unbranched Bur-reed and others.

How can these strikingly varied patterns be explained? Needing consideration are several intimately linked factors of which climate, soil and man's activities are the most important. The factor which overrides all others is climate. Some plants can tolerate temperatures below freezing and others cannot. The whole of Britain has a frosty climate: some very frost-tender plants cannot survive and regenerate in the wild year after year. Tomato plants spring up by the thousand in sewage sludge and can be found growing on shingle banks in the Clyde at Cambuslang and Low Blantyre. However Tomato will never spread to become an established member of the flora: the climate is too cold and the growing season too short.

Lack of water is not a seriously limiting factor for plants growing around Glasgow. Precipitation is plentiful and evenly spread throughout the year with a large number of rainy days. The mean annual precipitation during 1967–85 was 1,001 millimetres at Glasgow Airport, only four kilometres west of the western boundary. Since records began the wettest year was 1872 with 1,564 millimetres and the driest 1933 with 721 millimetres. Cloudiness and moderate temperatures mean that evaporation is not excessive. Climatically Glasgow has a transitional location. Only short distances to the north and west precipitation is much greater, cloudiness more, temperatures lower. Some plants may well be at or near their north-western climatic limits. The plants showing the south-eastern pattern may be examples.

The relationship between plant and soil is very important and highly complex. The physical properties matter greatly; depth and coarseness and fineness (stony, sandy, clayey) and *humus* (decaying remains of plants and animals) control the water-holding. Also of great importance is the chemical nature of the soil. The amount of available nitrogen, phosphorus, calcium, iron and other nutrients varies greatly from soil type to soil type. The acidity or alkalinity of the soil is crucial; this is called the *soil reaction*. Some plants are found mostly or entirely on very acid soils and not on mildly acid or alkaline soils that develop over limy rocks such as limestone. They can be called *lime-haters*. Conversely other plants cannot tolerate the very infertile, even toxic properties of very acid soils and thrive only on alkaline or mildly acid soils. They can be called *lime-lovers*. Many of the plants showing the southern pattern are lime-lovers. There may be greater availability of limy soils in the south; this needs investigation. The plants showing the north-eastern patterns are lime-haters. All the Glasgow area peat-bogs are in the north-east; the deep peat is very acidic and lacking in lime.

The L-shaped pattern is shown mostly by plants of broad-leaved woodlands growing on not too impoverished soils. There are woodlands in the north-east but many are coniferous plantations on poor soils such as the drained and drying-out peat-bogs.

So much heat escapes from the closely spaced buildings in cities that the temperature is raised, if only slightly, and especially so in winter. Slight as it is, it may well be enough to assist some of the *city dwellers* to survive especially alien invaders from southern Europe such as Oxford Ragwort. A spreading city changes the environment in a whole host of ways. Wet and marshy gound is inconvenient; cities are drier places than the country. Marsh plants are conspicuous among the *country cousins*. However, Mouse-ear Hawkweed is not at all a plant of marshy ground and with no less than thirteen other species it inhabits the flat roofs of the Physics and Astronomy Department of Glasgow University. Nevertheless it is a *country cousin* as Map 5 shows.

The *easy-goers* are plants that produce many widely dispersed seeds (or other propagules) and their habitat requirements are to be found readily in most or all of the squares.

Man creates new habitats like canals and railways. The narrow patterns in lines are those of plants such as Gypsywort and Lesser Water-parsnip following the canals or others such as Garden Angelica and Fennel Pondweed following the Clyde.

The full understanding of these and other patterns unmentioned here is a very complex matter and there are usually no clear-cut, simple answers. It is certain that not all the explanations lie only in the present or very recent past. Long-term changes, changes over at least the last 10,000–15,000 years, have great bearing on the patterns plants show today. Climatic fluctuations, soil maturation, vegetation changes and the use man has made of the land from the earliest times all have to be considered. Many plants well able to thrive in the environmental conditions of the Glasgow area could not have arrived by natural means because Britain was cut off from mainland Europe about 8,000 years ago. The outstanding example is Rhododendron. Though they can regenerate, Scots Pines do not grow indigenously around Glasgow now. That they did so several thousands of years ago between Easterhouse and Drumpellier Country Park is shown by the remains of stumps and logs preserved in peat. However at no time since the last Ice Age has central Scotland carried dense woods of Scots Pine, as studies of the microscopic fossils in peat and lake mud show. Figs of foreign origin are eaten by modern Glaswegians, Fig is very occasionally grown under glass and there is even a single wild Fig on the vertical stonework bank of the Clyde near Dalmarnock Bridge. How did it get there? The first inhabitants of the Glasgow area known to have eaten Figs were the Bearsden Romans, who in the second century AD also imported Coriander, Dill, Opium Poppy and, in large quantities, the Wheats, Emmer and Spelt. We shall never know if any Fig pip managed to escape and grow during the less than twenty years of occupation of the fort. It is very plausible, however, that seeds of such weeds as Corncockle and Soft-brome grew after spillage of the Roman grain, just as foreign weeds still grow, if only ephemerally, near the dockside granaries at Meadowside in Partick.

PART 2

Some Plants of Special Habitats

CHAPTER 5

Banks of the Clyde and Lochs

To the east the Clyde enters the area north of Blantyre and meanders through the rich Bothwell Woods, past Uddingston, past farms and disused pit bings. It goes over the weir at Carmyle before entering the more or less continuously built-up districts downstream of the now defunct iron and steel works that operated for 200 years. It loops round Cunningar and past Glasgow Green, acquired by the city as long ago as 1662. In the river downstream of Bothwell there is a diversity of water plants: Water Crowfoot, Shining Pondweed, Unbranched Bur-reed, Canadian and Nutall's Waterweeds. Even the heart of Glasgow has been reached by Broad-leaved Pondweed and Fennel Pondweed, the latter being known as somewhat tolerant of pollution. This is clear-cut evidence that the river's quality has much improved in recent decades, though far from being pollution-free. Past the tidal weir at the west end of Glasgow Green, the Clyde is a near-canalised river, the deepening of which made Glasgow's fortune. With vertical man-made banks, it flows much straighter past the now infilled and built over Queen's and Prince's Docks, through the ancient parish of Govan, where the main street had many thatched cottages until at least 1872. Passing the now much reduced shipyards, the Clyde leaves the area at Renfrew.

The principal tributaries of the Clyde from the south are the Rotten Calder and the White Cart Water. Those from the north are the North Calder and the Kelvin. Even in Kelvingrove Park, near where the Kelvin joins the Clyde, there are water plants such as Fennel Pondweed and Unbranched Bur-reed. This makes a very encouraging contrast with the sterile, smelly, frothy Kelvin of the 1950s.

Minor tributaries like the Camlachie, Tollcross, Yoker and Jenny's Burns and the famous Molendinar have long been partly or very largely out of sight, flowing through pipes. In the mid nineteenth century the Molendinar flowing through the historical heart of Glasgow was a foul, open sewer, a menace to public health. By contrast now, though they could scarcely be more urban, the short stretches of the Camlachie Burn above ground at the Gallowgate and at Carntyne Station support Water Plantain and Reed Sweet-grass. They would not do so if they were grossly polluted, though they are repositories of miscellaneous junk. At Nitshill, the Brock Burn, a tributary of the Levern, is a graveyard of dozens of supermarket trolleys. A very common component of our burns, even of the Allander in the middle of affluent Milngavie, these trolleys are very unsightly but not polluting.

Within the area there are ten lochs, all in the north. The smallest, both in Bearsden, and roughly circular, are St Germains and Kilmardinny. The latter is the larger but only about 250 metres in diameter. Reaching a length of about 1,000 metres the longest is Gadloch, near Lenzie, but being narrow it is exceeded in area by Bishop Loch at Easterhouse which covers about twenty hectares. Hogganfield, Bardowie and Bishop Lochs are much the same in size. All the lochs are shallow or very shallow and low-lying (below 100 metres above sea-level).

All the seven reservoirs are in the south and situated about 100 metres and more above sea-level. Close together at the south-west corner are Balgray, Glanderston, Littleton, Ryat Linn and Waulkmill Glen Reservoirs. With an area of sixty-five hectares Balgray is the largest standing body of water. Between Carmunnock and East Kilbride are Highflat and South Cathkin Reservoirs, the latter drained and out of use for many years.

Giant Hogweed (*Heracleum mantegazzianum*)

Growing up to three metres or more high (see Figure 1), this spectacular plant is a native of Caucasia where it grows in river valleys. It has escaped very successfully from gardens, as much in Glasgow as elsewhere. We have made records from twenty-seven squares, mostly central and southern.

The largest stands are on the banks of the Clyde, downstream of Carmyle, the Kelvin, downstream of Kelvinbridge and the Cart, downstream of Cathcart. The present abundance has arisen in the last half century. Although it had been recorded by 1950, as a student in the mid 1950s I cannot recall seeing Giant Hogweed in Glasgow. Now one of the most remarkable natural history sights of the Glasgow area is the massive stand growing out of the riverbank and wasteground of the Cunningar Loop. It can grow on road and motorway sides, as near the Kingston Bridge.

On the skin the juice from the stalks and leaves exposed to bright sunshine causes dermatitis which can be unpleasant. The effects have never killed anyone and probably never will. Nevertheless, much effort and money has been spent eradicating Giant Hogweed from parts of the country, as in Edinburgh. Under the Wildlife and Countryside Act 1981 it is an offence to grow Giant Hogweed in the wild.

PLATE 9

Dog Rose

× *0.9 natural size*

PLATE 9

PLATE 10

PLATE 10

Selfheal

× *1.0 natural size*

Garden Angelica (*Angelica archangelica*)

This sweet-smelling plant (see Figure 2) with culinary uses is a native of northern and eastern Europe. In the Glasgow area it was first found as a wild plant by Peter Macpherson in 1984. He encountered large numbers on both banks of the Kelvin where it flows into the Clyde. It is still thoroughly well established there but has now been recorded upstream to the Victoria Bridge and downstream to Yoker and Renfrew. It has also been found on the sides of the Thames, Trent and Mersey.

Hybrid Yellow Waterlily (*Nupar x spennerana*)

Two waterlilies, yellow (*Nuphar lutea*) and white (*Nymphaea alba*) are locally common in Scotland but both occur only sparingly in the Glasgow area, no longer growing in Hogganfield and Frankfield Lochs. A third, Least Waterlily (*Nuphar pumila*) grows no nearer Glasgow than Mugdock Loch. Even rarer is a hybrid between Yellow and Least Waterlilies with only about fifteen localities in the whole of Britain, mostly in Scotland. In the absence of the parents, Hybrid Yellow Waterlily (see Figure 3) grows in St Germains, Kilmardinny and Dougalston Lochs. At the first, it grows only in small amounts but at the latter two it is established in some profusion.

Mudwort (*Limosella aquatica*)

This small plant (see Figure 4) has a very peculiar lifestyle. As dormant seed it can survive in mud for a very long time. When the mud is exposed the plant grows briefly and quickly and dies after setting more seed. Well-known writers of guides to the

British flora have described Mudwort as extinct in Scotland. They should have seen the vast multitude of Mudwort at the dried-out Balgray Reservoir in 1984. Two years later, John Lyth found about ten plants on the banks of the Clyde at Cambuslang; soon after, this small colony was obliterated by a spate.

Few-flowered Leek (*Allium paradoxum*)

Like Giant Hogweed, this Leek (see Figure 5) is both a native of south-west Asia and an escape from our gardens. There may be only one or a few flowers per stem but there are groups of yellowish bulbils which drop off and spread the plant. It was first reported in the Glasgow area before 1930. In the last few decades it has increased a lot, now growing in fifteen of the squares. Locally abundant, it inhabits mainly riverbanks of the Clyde downstream from Bothwell to Dalmarnock and of the Kelvin downstream from Torrance to Kelvingrove.

Round-leaved Saxifrage (*Saxifraga rotundifolia*)

This Saxifrage (see Figure 7), an inhabitant of more or less shady streamsides, is a native of the mountains of central and southern Europe. Though never in the masses it forms in the Massif Central of France, it grows along the Kelvin from upstream of Killermont and downstream to Kelvingrove. For many decades it has been known near Lennoxtown. In the whole of Britain only at these places in and near Glasgow does it grow as a wild plant. Nevertheless it can be regarded as fully naturalised and should be recognised as such by botanical writers who, other than John Lee, have not hitherto mentioned it.

Indian Balsam (*Impatiens glandulifera*)

Very colourfully pink along the Clyde at Dalmarnock is Indian Balsam (see Figure 6), a native of the Himalayas. In 1933, John Lee wrote of it 'naturalised and growing profusely on the banks of the Kelvin within the boundaries of Glasgow'. Such abundance now applies to the Clyde and White Cart Water as well. Indian Balsam has been recorded from thirty-nine of the squares, many of them in the centre and south. If planted, this frost-tender plant, the tallest annual in the wild British flora, spreads itself around gardens by releasing its seeds explosively. The seeds need the freezing temperatures of winter and then they all germinate together in the spring. In gardens, Indian Balsam can be something of a pest but it is easily and cleanly pulled up from the soil.

Black Mustard (*Brassica nigra*)

Though it is occasionally found on wasteground, Black Mustard is very much at home along the banks of the Clyde upstream of Glasgow Green to Uddingston (see Figure 6). Here and there it is so well established as to colour the banks yellow, as at Dalmarnock. Black Mustard has an extensive geographical distribution but where it is native is uncertain. However, in the colourful abundance of Black Mustard, Dalmarnock resembles Galilee where, according to Michael Zohary, Black Mustard is a conspicuous part of the vegetation. In one other botanical way Dalmarnock resembles Galilee. There is a wild Fig tree: in the Holy Land there are many Fig trees, both wild and cultivated.

Fig (*Ficus carica*)

Only very rarely have Fig trees begun to grow on waste places in the Glasgow area. None is known at present except the Dalmarnock one, with thin multiple trunks more than three metres tall. Very remarkably it grows from the vertical stonework of the south-facing bank of the Clyde near the railway bridge downstream of the Dalmarnock Bridge (see Figure 6). Its position rules out having been planted. In any case planting is a very infrequent activity here, even under glass. Wild Figs may seem rather unlikely in Glasgow but they are far from unknown in more southerly cities, such as Dublin where a tree grew on the bank of the Royal Canal and Bristol where there is a tree again on vertical stonework of the Avon near Temple Meads Station.

Benefiting from the combination of south-eastern and urban warmth, Figs are not infrequent in London and Rodney Burton suggests that human faeces may have helped spread the plant. Certainly the pips can pass undamaged through the alimentary canal, as recoveries from many archaeological layers testify, not least the Roman sewage from Bearsden. Intriguingly the Dalmarnock tree grows near the sewage works but I was unaware of any Figs growing in sludge beds, which produce such great abundancies of Tomato plants. However, now it is known that along the River Don in Sheffield there are thirty-five or more Fig trees. One kilo of silt from that river produced fifteen plants of Tomato, six of Strawberry, one of Citrus and one of Fig. Sewage dispersed in water is now the probable source of these urban riverside Figs in Britain, including the Dalmarnock tree.

CHAPTER 6

Canals, Railways and Motorways

The first canal to be built in Scotland was the Forth and Clyde Canal which enters the Glasgow area at Kirkintilloch and leaves at Clydebank. Construction began in the east in 1769, Glasgow was reached in 1777 and the Clyde at Bowling in 1790. The Monkland Canal, made to bring coal from the Lanarkshire mines, joined the Forth and Clyde in the 1790s. The profits from the canals were considerable and the heyday spanned the years 1820 to 1840. Thereafter competition from the railways was fatal, even if it was a lingering death in the case of the Forth and Clyde. The canal finally closed to through traffic only in the early 1960s and considerable stretches are still open for leisure purposes.

A canal linking Glasgow to Paisley (the Paisley Canal) was built too late to be successful. By 1865 Roger Hennedy mentioned plants lost when the canal was converted to the railway which still exists. The destruction of a long stretch of the Monkland Canal came in the 1960s when it was drained and filled to become part of the M8 motorway. In 1955 while hunting plants along the towpath at Ruchazie I found the rare, lime-loving Quaking Grass; this was its only occurrence within the Glasgow area. The building of the M8 must have destroyed this attractive Grass.

PLATE 11
The Cathkin Braes in late June

× 0.4 natural size

At the south of the Glasgow area is the steep north-facing rocky slope of the Cathkin Braes. From the public park at almost 200 metres above sea-level, there is a fine panorama of the now smokeless city with Castlemilk in the immediate foreground and the Campsie Fells and Kilpatrick Hills in the distance.

Like Possil Marsh, in the nineteenth century the Cathkin Braes were a favourite haunt of botanists who found plants which cannot be found now. Thomas Hopkirk knew Alpine-Clubmoss 'on the top of the Cathkin hills', Petty Whin 'on the moors on Cathkin hills', Fragrant Orchid 'on Cathkin hills immediately above Castlemilk' and Small-white Orchid 'in a pasture above Castlemilk'. The distinguished William Jackson Hooker found Common Rock-rose and Roger Hennedy listed Fir Clubmoss.

A friend of Roger Hennedy, Hugh Macdonald was a keen natural historian and a successful journalist. In 1854 he published the very successful Rambles Round Glasgow *in which he has this to say in his flowery style about the Cathkin Braes:*

> *Between the summit of the braes and Carmunnock, about a quarter of a mile to the southward of the road and on a wild tract of moorland, are the traces of an ancient British camp. To this spot we now direct our steps, disturbing on our way several snipes, which have bred among the moist, marshy hollows. . . . The tufted cannach [Gaelic for Hare's-tail Cottongrass] here waves in the blast its snowy plumes, the curious sun-dew* (Drosera rotundifolia) *is also met here, with its glittering beads of dew unmelting 'in very presence of the regal sun', with marsh violet* (Viola palustris)

> *creeping in beauty along the untrodden heath, and the buckbean* (Menyanthes trifoliata) *and marsh cinque-foil* (Comarum palustre) *rising above the dark moss-water.*

In the 1910 edition of the Rambles, *Rev G H Morrison states in a laconic footnote about the wild tract of moorland walked over by Hugh Macdonald: 'Where now is the golf-course'. It sounds like a disastrous transformation concerning rare, wild plants of bogs. Neither Sundew nor Bogbean are there now but others have survived in the remaining small boggy, heathy parts of the golf course and immediately adjacent land. Pill, Spring and Green-ribbed Sedges, Cranberry, the two Cotton-grasses, Marsh Violet, Marsh Cinquefoil and Cross-leaved Heath are all still there.*

The plants shown in Plate 11 come from the diversity of habitats that still persist, if only precariously in some cases, between Castlemilk, Carmunnock and Highflat and South Cathkin Farms. Sensitive management is needed if the diversity and ecological interest is to be maintained. Germander Speedwell, with blue flowers, is to the bottom right. To the bottom left is Bitter Vetch with pods. Middle left are the dangling heads of Green-ribbed Sedge and middle right the small white flowers and finely divided leaves of Pignut. The flowering heads centre top are those of the Downy Oat-grass and the three whitish heads are Hare's-tail Cotton-grass.

PLATE 11

Figure 7 *Round-leaved Saxifrage at Garscube.*

Figure 8 *Dark Mullein along railway at Anniesland.*

Figure 9 *Cowslip on railway cutting at Arkleston.*

Figure 10 *Oxford Ragwort at Bunhouse Road, Partick.*

Figure 11 *Butterfly-bush at High Street.*

Figure 12 *Royal Fern at Port Dundas.*

Figure 13 *Austrian Yellow-cress at Possilpark.*

Figure 14 *Sheep's-bit (blue flowers) at Kirkintilloch with Cat's-ear (yellow flowers).*

Water flows through the Forth and Clyde Canal westwards at the volume of a few to over twenty million litres a day; the canal is far from stagnant and nor is it anywhere so polluted as to kill off all the plants.

Inhabiting the canal are a diversity of submerged aquatic plants such as Horned Pondweed and Rigid Hornwort, both in the Glasgow area confined to the canal. Bennett's Pondweed, easily found in the canal, is a hybrid confined to central Scotland. Fen plants take advantage of the sides of the canal; Reed Sweet-grass occurs in very considerable abundance.

The first railway with steam trains in the Glasgow area was that from the city to Garnkirk, opened in 1831. As the city flourished as a major industrial centre an extensive network of lines was soon laid out. That some plants can take advantage of the varied, largely competition-free conditions provided by the construction of railways was known to Roger Hennedy before 1865. He knew Common Toadflax 'on a railway bank near Eastfield' (Rutherglen) and Sticky Groundsel 'on a railway running fron the west end of Cadder Wilderness to the Forth and Clyde Canal, plentiful'. By the 1891 edition of Roger Hennedy's book, Field Bindweed was known to be abundant at Parkhead Station. These three plants and many others have found railways very congenial places in which to grow.

Roger Hennedy did not just record the colonists of railways, he told of a disappearance. By 1865 there was only one locality of Lesser Skullcap in the Glasgow area. This was Rosebank on the Clyde near Rutherglen, 'now cut up by a railway'. Lesser Skullcap has not been seen again in the Glasgow area. It is the only plant known to have been rendered extinct in the area by railway construction.

Common Toadflax, Sticky Groundsel and Field Bindweed are

far from being uncommon around Glasgow. However, many unusual plants are often found along railways. Stations and nearby land are happy hunting grounds for the botanist. In the mid 1980s Hyndland Station was such a place. Table 1 is a list of its flora. More than half of the plants are among the commonest in the Glasgow area and indeed much of Britain. However no less than eight are rare in the Glasgow area. Isle of Man Cabbage is one of the very few *endemic* plants in Britain; *endemic* means restricted to a particular area. It is a seashore plant found nowhere but around the British coasts of the Irish Sea and the Firth of Clyde. It has occasionally been found away from these coasts but always in places much disturbed by man. How it came to grow at Hyndland Station and elsewhere in the Glasgow area (Dawsholm, Sandymount and Gateside Bing, along or near railways) is an interesting puzzle. Somehow railway traffic has brought it from the coast, as is the case with Sand Sedge at High Street and Sandymount. A plant with long-lived seeds, Henbane is very seldom found in the Glasgow area. At Hyndland Station there was only one plant. Swamp Meadow-grass is rare in Britain as a whole and very rare in Scotland. Sometimes it lives up to its name but it can grow in well-drained bare ground as at Hyndland Station. By 1990, if not before, seven of the eight had vanished. No flood of tears is appropriate. The seven are all plants that would disappear if the ground were left undisturbed. They thrive only on unshaded, disturbed and bare ground. Closing up of the vegetation would eliminate them. The eighth is the very unusual hybrid Rowan discussed in the next section.

Over many years as railways have become unprofitable, so they have been closed down. In the Glasgow area about sixty-five kilometres of disused lines and immediately adjacent active lines have been surveyed, mostly by Henry Noltie and Keith Watson and the MSC team. Many uncommon and rare plants, including some that could scarcely have been predicted, have been discovered. They include Common Restharrow, Danish Scurvy-grass, Knotted Pearlwort, Lamb's Lettuce, Little Mouse-ear, Round-leaved Sundew, Royal Fern, Sand Sedge, Sea Mouse-ear, Stag's-horn Clubmoss, Twiggy Mullein and Water Chickweed.

The M8 motorway zigzags across the Glasgow area from Bargeddie in the east to Renfrew in the west. The eastern half was surveyed by the MSC team who had in mind salt-tolerant plants. It had already been well established that seashore plants such as Annual Sea-blite, Saltmarsh Rush, Sea Aster and Sea Plantain were spreading along motorways in England. The use of salt as a de-icer makes the verges of the motorways salty and so provides a suitable habitat. However no such plants appear to have arrived in the Glasgow area, unless Foxtail Barley is counted. This ornamental alien from North America is somewhat salt-tolerant. For several years in the 1980s it grew along a slip road joining the M8 at Alexandra Park. However, the M8 is far from lacking in botanical interest. Some examples of uncommon or otherwise unexpected plants are Common Centaury, Hemp Agrimony, Wild Parsnip and Lucerne.

Cowslip (*Primula veris*)

With a mainly southern distribution in Britain, this lover of limy soil is uncommon in Scotland. Thomas Hopkirk realised a long time ago that it is not indigenous around Glasgow, where it is known only from a few scattered places, mostly very near gardens. Hybrids with Primrose have been reported from Bothwell where Cowslip has been known for nearly 200 years.

There is only one place where Cowslip occurs in large numbers (see Figure 9). This is the south-facing slope of the railway cutting at Arkleston, well away from any garden.

Dark Mullein
(*Verbascum nigrum*)

In Scotland, and certainly in the Glasgow area, the Mulleins are strongly connected with railways. Great Mullein, the only one well established in Scotland is discussed in Chapter 8. Twiggy Mullein, which has hardly ever been recorded in Scotland, was found on railway ballast at Carmyle by Keith Watson, and Dark Mullein (see Figure 8) was found in 1989 in a similar habitat, a little north of Anniesland Station. Dark Mullein has also been found on the south-facing slope of Waterside Bing. Both Twiggy and Dark Mullein share very southern ranges in Britain and may find difficulty in persisting in Scotland. However, John Lee recorded Dark Mullein at Mugdock Castle more than fifty years ago. Though no longer at the castle, it still grows within Mugdock Country Park.

Prickly Heath
(*Pernettya mucronata*)

A native of Chile, the cultivated Prickly Heath is a small shrub with attractive berries which are spread by birds. There are occasional reports of wild plants from various parts of Scotland. One bush is spreading on disused, flooded railway sidings near the city centre (see Plate 2) and it is naturalised at the Linn Cemetery on a north-west-facing slope.

Common Spotted Orchid
(*Dactylorhiza fuchsii*)

This Orchid (see Plate 3) has spotted leaves and is common in Britain and so lives up to its name. It is by far the commonest Orchid in the Glasgow area with records from no less than seventy-six of the ninety squares. The habitats are various: coal bings, lime bings (Milngavie), roadsides including motorway embankments, disused railways, the rubble-infilled Monkland Canal near Swinton, undisturbed wasteground, pasture and marshy places.

Singletons can occur almost anywhere: at the foot of a factory wall at Lancefield Street, Finnieston; among gravestones at St Peter's Cemetery, London Road; the demolished Dalmarnock Generating Station; an abandoned railway at the closed Parkhead Steelworks now replaced by a large shopping centre, a roadside at Port Dundas and the abandoned walled garden at Bothwell Castle.

It can occur in hundreds as on flooded, disused sidings near the city centre where it grows mixed with Northern Marsh Orchid.

Northern Marsh Orchid
(*Dactylorhiza purpurella*)

The habitats of Northern Marsh (see Plate 4) and Common Spotted Orchids are broadly similar though the former is less than half as frequent as the latter. Not infrequently, they grow together and hybridise; the hybrid grows in rooftop garden tubs in Hyndland.

Common Reed
(*Phragmites australis*)

Common Reed (see Plate 1) grows in only five of the Glasgow squares, all in the northern half. Its large, creeping, underground stems often grow into extensive patches as at Possil Marsh, Bishop and Johnston Lochs. Preserved in fen peat, these stems stay recognisable for thousands of years. On the banks of the Clyde near Renfrew, there is only a small patch and none all the way upstream to Blantyre. Perhaps as a deliberate planting it grows at Gartshore Estate, east of Kirkintilloch. Despite growing at Possil Marsh immediately alongside the canal, Common Reed has not colonised the banks of the canal there or elsewhere in the Glasgow area; seed set is known to be often very low and germination very poor in the wild.

Common Reed has a diversity of uses. Harvested in the Danube delta, it is used to make paper, cellophane, cardboard and synthetic textiles, fibreboard, fuel, alcohol, insulation and fertiliser. The leaves can be made into mats and the grains and young shoots eaten. All this is listed in the very useful dictionary by David Mabberley who does not mention its well-known use as durable thatching. As the photographs in Maurice Lindsay's book clearly show, there were thatched roofs in Glasgow and surrounding areas up to the end of the last century. What was used for thatching? The nineteenth-century botanists mention House Leek, Climbing Corydalis and even Sea Mouse-ear as growing on thatched roofs but do not mention so mundane a thing as of what the thatch was made. Perhaps Reed was used but more likely the Glasgow cottages had Oat or Barley straw or Heather, these being more readily available than Reed.

Great Reedmace (*Typha latifolia*)

Thomas Hopkirk knew of only one locality for this tall conspicuous plant (see Plate 1). Roger Hennedy called it very rare. John Lee considered it to be 'frequent'. Now we know it in thirty-two of the Glasgow squares. A spread since the middle of the last century seems to have taken place. It is not clear why this should be, though around London a similar spread has been ascribed by Rodney Burton to freedom from grazing by cattle no longer visiting ponds for watering.

Great Reedmace is a plant of loch shores, fens, slow-moving water and even small areas of marshy ground with very little depth of standing water, as in the Garngad and in Carntyne on wasteground, now built over.

Gypsywort (*Lycopus europaeus*)

In Britain as a whole, Gypsywort (see Plate 1) is common only in the south. In Scotland it is uncommon and largely coastal. Around Glasgow it is known from twenty of the squares, all in the north. The habitats are watersides, especially the canals and fens.

The absence of Gypsywort from the whole southern half of the area and from the banks of the Clyde and its tributaries is very striking. The distribution pattern coincides with the lines of the Forth and Clyde Canal and the defunct Monkland Canal. No author recorded Gypsywort near Glasgow before 1865 when Roger Hennedy stated baldly: 'Forth and Clyde Canal'. It is tempting to think that the plant invaded along the canals. The localities furthest away from the canals are only about three kilometres or less distant as at Dougalston Loch.

The canal invasion theory needs to take account of two fruits of Gypsywort recovered from the Roman sewage-filled ditch at Bearsden. How did they become entombed in the smelly, silty clay if Gypsywort did not grow around Glasgow till over 1,700 years later? Perhaps the Romans had brought it for one or other of its useful properties: dyeing and medicine.

Rosebay Willowherb (*Chamaenerion augustifolium*)

Thomas Hopkirk knew this handsome plant (see Plate 1) but only 'about the banks of the Clyde at Barncluith'. Rev Patrick also knew it at Barncluith (which is on the Avon, not the Clyde) and 'in great abundance' on the north-east bank of the Clyde 'a little below Hamilton bridge'. In 1865 Roger Hennedy considered it 'an escape from cultivation' and listed only 'Stonelaw Wood'. As a young man of eighteen, I vividly remember being told by John Lee, nearing ninety, that in the late nineteenth century the native Rosebay Willowherb was a rare mountain plant. In his book John Lee states that in low-lying ground 'It is an introduction but has spread all over the country in recent years and has now become a feature of open ground, railway banks, etc.'

Rosebay Willowherb is found everywhere in the Glasgow area in many different habitats: gardens, roadsides, edges of woods, railway embankments, wasteground, including bings of lime (Milngavie) and coal. It is one of the most conspicuous colonists of derelict buildings in the city centre.

The great spreading of Rosebay Willowherb in low-lying parts of Britain during the last 150 years or so and especially in the early twentieth century, has been discussed many times, most recently and cogently by Oliver Rackham. He accepts that lowland plants are aliens and claims that Lowland Rosebay was probably an introduction into England from America or Central Europe in the seventeenth century.

Royal Fern (*Osmunda regalis*)

Nobody had ever recorded this handsome, large Fern (see Figure 12) growing wild in Glasgow until a few years ago. Now we know of it in no fewer than three places. It would not have survived for long had it been present in Victorian times. Here is a report published nearly 100 years ago by Thomas King and D A Boyd.

> Hunterston, Ayrshire; formerly luxuriant on the cliffs and low ground facing the sea, but now extinct, the plants having been carried away . . . Portincross, Ayrshire; a few plants formerly grew here but were all dug up and carried away . . . Shewalton Moss, Dundonald, Ayrshire; extinct the moss being reclaimed . . . Island of Cumbrae, Buteshire; formerly plentiful in places, this fern has almost entirely disappeared being carried away . . . Island of Arran, Buteshire; up till 1860 abundant in many places, but now extinct, or nearly so, having been carried away by the cartload and boatload . . . Roots are still sometimes offered for sale to visitors to the island . . . Achacha District, Benderloch, Argyllshire; formerly plentiful but none have been seen for twelve to fourteen years, the roots having been continually sought after and removed by collectors . . . Loch Fyne District, Argyllshire; becoming extremely rare through ruthless collections for sale . . .

The Victorian craze for growing Ferns both in and out of doors has been written about in a book by David Allen and the standard account of British plants states about Royal Fern 'now almost extinct in most heavily populated areas owing to depredations of collectors'.

Familiar with the decline of Royal Fern, I was very reluctant to believe a report by Basil Bush that there was one plant on the bank of the canal. Soon after, no less than about fifty plants were found on the vertical stonework banks of the canal at a different place by Keith Watson and the MSC team. Only in 1989 was another singleton found on flooded overgrown railway sidings near the city centre.

Ferns spread by structures called spores which, being microscopic, can be blown over large distances. Where did the colonising spores that developed into the wild plants in Glasgow come from? You can see just how striking and massive a Fern this is in the Kibble Palace of the Botanic Garden. Perhaps the spores left this magnificent cultivated specimen and began the spread in Glasgow or maybe the spores came from wild plants around the Clyde coasts or from the southern Inner Hebrides where Royal Fern is not too rare. Glasgow is not alone in supporting urban wild plants of Royal Fern. There are reports from Lancashire and London.

CHAPTER 7

Rubbly Wasteground

Old and even modern housing may need replacing, businesses and industries fail and the buildings are demolished. Recent years have seen the final passing of much of the heavy industry in Glasgow. The constant need for redevelopment in big cities means that there are created areas of rubbly wasteground, some large, some small, that lie vacant, sometimes briefly, sometimes for several years or more. These places are immediately colonised by plants, even though the rooting conditions may be very poor.

Expanses of concrete and tarmac provide little foothold. Piles of broken brick, cement, plaster and lumps of concrete are very freely draining. There may be acute deficiency of nitrogen, crucial for plant growth, and little or no organic matter to hold water and release nutrients. However, some important nutrients such as phosphorus and calcium may be in good supply in the rubble and the soil reaction favourably high.

At the heart of the old city, the area bounded by High Street, Bell Street, Hunter Street and Duke Street was in the late 1980s mostly rubbly wasteground with two active railway lines and High Street Station. There were stretches of redundant cobbles, exposed walls of former cellars and other remains of totally derelict buildings. Those fourteen-plus hectares have had a chequered, interesting history. In the sixteenth to eighteenth centuries and for much of the nineteenth century, a large part of the area was covered by the buildings and gardens of Glasgow University. Over many years, the university authorities carried out a great deal of planting of trees and other plants. In the late seventeenth century for example, Apple, Cherry and Pear trees were planted, in 1704–5 Horse Chestnut, Lime and Plum trees, in 1706 no less than 1,150 Beeches and in 1725, 200 Lime trees. Red and White Clovers and Ryegrass were sown in 1728 and in 1705 a garden for medicinal plants was laid out. It lasted until after 1803 when Faculty Minutes stated that 'the smoke from the Foundery was very prejudicial to the Botany Garden.' A type foundry had been set up in 1762 and released fumes leading to heavy metal pollution. All this and more has been very readably related by Don Boney in his book, *The Lost Gardens of Glasgow University*. By 1870 the foul, smoky air and the slums of High Street were too much and the university moved to the more salubrious west at Gilmorehill. The ground was sold and eventually became a railway; the College Goods Yard lasted until 1968. Now the area is being redeveloped as a large car park and new housing is being completed and the former bonded warehouse at Bell Street has been converted to flats.

As wild black rabbits darted away, I walked over the area in the summer of 1986 and recorded the plants I found. The results, with additions made on revisits in 1988, 1989, 1990 and 1991 are given in Table 2. At least 115 species had colonised the area by that last date, though some may have been already exterminated by the redevelopment. This richness of plants is not atypical of extensive patches of wasteground left undisturbed over some years.

Thirteen of the colonists are shrubs or trees including Silver and Downy Birches, Wych Elm and Sycamore, capable of growing into woodland. Free from too persistent interference by redevelopment, the vegetation would gradually and inexorably change from the initial stage of bare, rubbly ground with low-growing annuals to denser vegetation of taller perennials and finally to shade-casting woodland. This is what ecologists call *succession*. Under our climate on dry ground, succession leads to woodland. The vegetation of the High Street wasteground is at a very early stage of turning into woodland, the end-point (or *climax*) which will never be reached because of the redevelopment.

Some of the rarer plants in the list are very noteworthy. The famous, poisonous Hemlock has been recorded from only eleven squares during the recent survey; the unfamiliar grass, Rough Bent, from only nine, and Sand Sedge from only two.

Never in large numbers, Hemlock grows on roadsides, rubbish dumps and wasteground. It is a biennial, germinating one year and fruiting then dying the next year. At a roadside in Dennistoun, it has persisted at precisely the same spot for over thirty years. In 1756 Hemlock was a weed in the gardens of the university. During my visits to the rubbly wasteground in 1988 to 1990 I found Hemlock not far from the position of the former entrance to the university. Can it be possible that Hemlock has grown, if only intermittently, for nearly 250 years? Did the recent disturbances bring long-dormant seed to the surface? The Danish botanist Soren Odum has provided strong circumstantial evidence that Hemlock seeds stay viable for 150 years or more; the plant grew from soil taken from beneath a demolished house. Or has the small stand of Hemlock grown merely and disappointingly as a result of recent colonisation?

Little known in Britain and very much an urban plant,

Rough Bent has certainly not lain dormant for many years. It is a recent invader from its native North America. First found in Glasgow at the docks in 1973 by Peter Macpherson and Alan Stirling, it now grows on wasteground in various parts of the city. At the northern part of the High Street wasteground it grew in vast numbers in 1989 and 1990.

A well-named coastal plant, Sand Sedge has only very seldom been found inland in Scotland. In the Glasgow area it was unknown until 1989. Among the ballast along the active railway between Shettleston and Garrowhill Stations it grows and fruits well. There is a smaller and less vigorous colony along a low, south-facing wall above High Street Station. In addition to Rough Bent in the list from High Street there are seventeen other plants alien in Britain. It is characteristic of urban vegetation to find many alien plants and it is also typical of that vegetation for the aliens to be among the more successful colonists. Five of the thirteen most abundant species at High Street are aliens.

Of the following nine illustrated species six are aliens.

Austrian Yellow-cress (*Rorippa austriaca*)

Four species of Yellow-cress have been recorded in the Glasgow area during the 1980s. Two are easily found: these are Marsh and Creeping Yellow-cresses which inhabit gardens, nurseries in the parks and wasteground apart from riversides and pond verges. Great Yellow-cress is rare, being known from only two places, Newlands and Dalmarnock. Austrian Yellow-cress (see Figure 13) has only one place but there it occurs in some profusion. On rubbly wasteground, between Bilsland Drive and Balmore Road, Possilpark, perhaps there are more plants than anywhere in Britain of a species listed by Frank Perring and Lynne Farrel in the *Red Data Book* as known from only eight English counties.

Oxford Ragwort (*Senecio squalidus*)

Groundsel (*Senecio vulgaris*)

The genus *Senecio* has about 1,500 species showing great diversity in shape. There are bizarre rosette trees, which grow to five and a half metres tall and are peculiar to the East African mountains, as well as the familiar Groundsel, often only fifteen centimetres or less tall, growing in almost everyone's garden.

The flower-heads of Groundsel are usually very drab but sometimes they have prominent yellow petals. It is plants in the streets and wasteground of Glasgow that have the conspicuous petals. Some botanists think that these conspicuous petals have arisen because of crossing with Oxford Ragwort. Others are not so sure. It is a controversial matter. Oxford Ragwort is a native of southern Europe, not of Oxford where it has been grown in the Botanic Gardens since the late seventeenth century. However, not until this century has it spread rapidly through Britain. In Glasgow the Oxford Ragwort is now known from fifty-two of the squares, many of them central and urban. As the photograph shows (see Figure 10), Oxford Ragwort is a plant of well-drained sunny places, such as south-facing walls and rubbly wasteground.

Butterfly-bush (*Buddleja davidii*)

This often cultivated shrub (see Figure 11), a native of China, is very attractive to butterflies. Like Oxford Ragwort, it inhabits

PLATE 12
Goldenrod
× *0.6 natural size*

PLATE 12

Figure 15 *Young's Helleborine on wooded bing.*

Figure 16 *Young's Helleborine closeup.*

Figure 17 *Dune Helleborine on wooded bing.*

Figure 18 *Dune Helleborine closeup.*

PLATE 13

PLATE 13
Garscadden Wood in mid May
× 0.4 natural size

One kilometre long, Garscadden Wood is roughly tadpole-shaped, with the head to the north-west and the tail to the south-east. It is an ancient wood, the head of the tadpole appearing on the old maps at least as far back as the mid eighteenth century, perhaps even as early as the late sixteenth century.

The tail of the tadpole grew later and had been planted latterly with alien Larch and Spruce and native Scots Pine. There was a lot of Sycamore, a tree brought to Britain before the seventeenth century and now fully naturalised. The head is of much greater interest with much Oak and Hazel, both regenerating freely. Blown down in the great gale of January 1968, many of the Oaks have sprouted vigorously from the more or less horizontal trunks. Running downhill there are ridges and furrows of the strips of the long-outmoded runrig cultivation. There are other earthworks that may be a sign of ancient woodsmanship.

There are other trees in the wood such as Ash and Wych Elm, though not in large numbers. In the Plate, the twigs are those of Wych Elm; that on the right carries as yet unripe, winged fruits which finally blow around freely. Young saplings of Wych Elm are easy to find in the Glasgow area but never those of the planted alien Wheatley Elm which does not set seed. Dutch Elm disease can be found wherever one goes in the Glasgow area and Garscadden Wood is no exception. This disease, which has had such sweeping effects in England is not so devastating in Scotland. However, Kenmuir Wood, a little upstream of Carmyle, has been badly affected. It is another ancient wood, formerly the haunt of Wood Vetch and Wood Fescue, both now extinct in the Glasgow area, and still the habitat of Moschatel and Meadow Saxifrage, as revealed by Agnes Walker and her helpers. The disease is not always fatal as can be seen at the tip of the tail of Garscadden

Wood. There some Wych Elms are diseased but still alive, with dead crowns but sprouting bases.

At the bottom of the Plate is the pink-flowered Pink Purslane which only grows in small quantity in the wood. Bottom left is Opposite-leaved Golden Saxifrage which grows along the tiny burns. The expanding fronds of two Ferns are Scaly Male-fern (centre) and Broad Buckler-fern (centre right). Centre left are the nodding, tubular flowers of Bluebell in the usual purple-blue colour and also in the less common white and pinkish white.

Garscadden Wood is situated on a south-west-facing slope less than 100 metres from north-eastern Drumchapel. On a walk through this colourful woodland in spring it is easy to meet local residents, young and old, who are interested in natural history. However, the wood is put to a variety of uses, many very detrimental: a place to set on fire, to dump rubbish, to burn out cars, a source of timber. Many trees have been cut down, mostly at about one metre above the ground. The wood deserves better at the hands of the residents. It is a place to cherish, not to despoil. Early in 1990 the Kilpatricks Project in collaboration with Glasgow District Council began to tidy up the wood and its immediate surroundings. One of the first actions was to cut down most but not all of the Sycamores; many of the cut stumps are sprouting again, the typical behaviour of many coppiced trees.

heaps of rubble and other well-drained places. Between High Street, Bell Street and Duke Street, the site of the demolished College Goods Yard, in the late 1980s there was a massive stand of Butterfly-bush, now infilled. It has colonised nearby buildings as in Bell Street and the streets running south into Trongate. There are now records from nineteen of the Glasgow squares, mostly central and all but one urban.

It is not only on the central rubbly wasteground and buildings of Glasgow that Butterfly-bush thrives. Rodney Burton tells us that 'in central London it is the commonest shrub' and for Bristol, Oliver Gilbert relates: '*Buddleia* dominates the city centre forming thickets everywhere. A slogan on a wall reads *Buddleia* rules OK!' According to Peter Jackson and Micheline Skeffington, Butterfly-bush is abundant in inner Dublin.

Rowan
(*Sorbus aucuparia*)

The native Rowan (see Plate 5) ascends our hills to a greater height than any other tree. It can grow on very poor soils and is shade-tolerant. It readily regenerates in parks and gardens, on wasteground and on drying-out deep peat. Only three squares lack records of Rowan.

Common Whitebeam (*Sorbus aria*)

A native of southernmost Britain where it grows on limy rocks or soil, Common Whitebeam (see Plate 5) colonises wasteground and places like Rowan. It is known from forty-two squares with a central concentration.

Swedish Whitebeam (*Sorbus intermedia*)

A native of northern Europe, Swedish Whitebeam (see Plate 5) colonises similar places to those of Common Whitebeam but is considerably the more frequent of the two. It has been recorded from sixty squares, strongly central and urban.

Pineappleweed (*Matricaria matricarioides*)

Crush the flower-heads and the pleasant aroma explains the name of this plant (see Plate 6) now known from eighty-nine of the squares. Its habitats include wasteground, roadsides, edges of arable fields and tracks. It is a native of north-eastern Asia and perhaps north-western America. Whereas the spread of Oxford Ragwort is often related to the railways, that of Pineappleweed is related to cars and roads. It is a plant of the more fertile soils and can form a persistent bank of seeds in the soil.

Dame's-violet (*Hesperis matronalis*)

This native of Europe and west and central Asia (see Plate 7) was little known outside gardens to nineteenth-century authors. John Lee wrote in 1933 that it was 'becoming naturalised in many places'. There are now records from seventy of the squares. It can sometimes be abundant, as on riverbanks and wasteground.

CHAPTER 8

Bings and Coups

B*ing*, a Nordic word meaning heap, is familiar to many Scots but very unfamiliar to many English. Another word, known to some Scots but not English, is *coup*; pronounced cowp, it is of French origin. It means a rubbish tip.

Bing is most often used as in the phrase coal bing, referring to the spoil heaps from mining. Most of the bings remaining in the Glasgow area are the products of deep mining for coal. However, there are bings from other types of mining as well as from industrial processes.

Ironstone Bings. North of Possil Marsh there are three small bings resulting from ironstone mining. One supports an unexpected grass, Viviparous Fescue, normally a plant of hilly ground. Found only in 1989 by Lorna Smith, it has never before been seen in the Glasgow area.

Steel Works Bings. A little to the east of Cambuslang at the demolished Hallside Steel Works there are low dumps of slag and extensive rubbly wasteground. Some of the very colourful plants that grow there are illustrated in Plates 8, 9 and 10. There is a single clump of the alien North American False Fox Sedge, very seldom encountered in Britain.

Paper Works Bings. At Milngavie, there are low mounds of carbonates, especially calcium carbonate (lime) that are the dumped waste product of a paper works. Now covered naturally by well-grown Birches, Willows and other trees, there is a remarkable suite of herbaceous plants including the carnivorous Butterwort and three Orchids, Broad-leaved Helleborine, Common Spotted Orchid and Twayblade.

Coal Bings. Though coal mining began in the Clyde Valley at least as early as the sixteenth century and was very extensive in and around Glasgow last century and well into this century, the only bings remaining are all to the east of the city. They derive from mining in the last 125 years. Thirteen bings have been searched for plants. In the north-east are Waterside, the only conical bing left, Auchengeich and Wester Auchengeich (removed by 1985) and Cardowan where is a complex of bings, one being on fire internally. Cardowan Colliery was the last to finish, in 1983, only a year after the closure of Bardykes Colliery in the south-east. Also in the south-east are the bings called Bothwell Castle, Dechmont 1/2 and 3/4, Gateside, Hallside, Kenmuirhill and Newton 1 and 2.

Bings, especially coal bings, can be very difficult habitats for the growth of plants. Varying from very blocky to fine grained, the spoil is loose and may be unstable. It may be toxic, having a very acidic reaction, produced by iron pyrites turning into sulphuric acid. By contrast some spoil has an alkaline reaction. It may be too dry, the drainage being unimpeded and those steep, dark-coloured slopes getting most sun may become exceedingly warm. As well as all these drawbacks, at first there is no organic matter in the spoil and there may be acute deficiencies of the crucial nutrients, nitrogen and phosphorus.

Despite these very hindering stringencies, plants do colonise coal bings. Not a few are fascinating because they are geographically unexpected, ecologically surprising or even because they are in the process of evolving.

Viper's Bugloss
(*Echium vulgare*)

Buglosses are mostly plants of sunny, warm climates. They are well represented in southern Europe and there are more than twenty species on the Canary Islands where some thrive on lava flows. In Scotland we have only one, Viper's Bugloss (see Plate 8), a strikingly handsome but uncommon plant which has a very southern occurrence in Britain as a whole.

In the Glasgow area Viper's Bugloss grows in fifteen squares and gets the sunny, well-drained, competition-free conditions it needs on bare ground, often bings, as realised long ago by John Lee. It grew abundantly on the coal bing at Wester Auchengeich in 1984. By 1985 it had vanished with the bing. It is known from three other coal bings: Bardykes, Gateside and Kenmuirhill.

However, in 1989 on the bings and surrounding rubbly wasteground at the demolished Hallside Steelworks it grew in such great profusion as to be a stunning display. The heaps of solidified slag, visually and even chemically like lava flows and supporting masses of purply-blue Viper's Bugloss, were very reminiscent of Tenerife.

Weld (*Reseda luteola*)

Wild Mignonette (*Reseda lutea*)

In the past Weld (see Plate 8) was cultivated for a yellow dye, as discussed by Su Grierson; an alternative name is Dyer's Rocket. Weld was first reported in the Glasgow area 200 years ago by

David Ure who knew it at the 'East-quarry, Rutherglen'. It has been recorded from fifty-six squares and its relative, Wild Mignonette, has been found in forty-two. With undivided leaves, Weld is the taller and less branched of the two. Both inhabit well-drained, sunny, bare wasteground, just like Viper's Bugloss. They often grow together, with Weld usually in the greater numbers.

Weld grows on eleven of the coal bings, being unrecorded only from Auchengeich and Wester Auchengeich. On one of the Cardowan bings it was very abundant. Recorded from eight of the coal bings, Wild Mignonette occurs in great numbers with Viper's Bugloss at Hallside Steelworks where Weld was a very minor presence.

Great Mullein
(*Verbascum thapsus*)

In Britain Great Mullein (see Plate 8) is common in the south while in Scotland it is a lowland plant. In the Glasgow area it has been found in twenty-four squares, fifteen of which are central.

Like Viper's Bugloss, Weld and Wild Mignonette, it is a plant of bare, well-drained, sunny places. Great Mullein has been seen on the coal bings at Waterside and at Hallside Steelworks. It is strongly connected with railways, both active and disused.

Earlier authors, even Robert Grierson, seem not to have found Great Mullein very much in the Glasgow area. Perhaps there has been a spread in the last half century.

Woody Nightshade (*Solanum dulcamara*)

A scrambling plant, sometimes found in large masses as on the banks of the canal near Temple, Woody Nightshade (see Plate 8) has been found in over sixty of the squares. Two hundred years ago David Ure knew it in hedges near Rutherglen. Nearly 2,000 years ago a pollen grain fell into one of the ditches round the Roman fort at Bearsden.

It is often found on wasteground – even at the heart of the city, near Queen's Street Station. There are only a few records from bings.

This native is a poisonous plant with red berries. It is not so toxic as the dark-berried Deadly Nightshade which has only very seldom been found in and around Glasgow.

Large-toothed Hawkweed (*Hieracium grandidens*)

Spreading through Britain, this Hawkweed (see Plate 8) is a native of the mountains of central Europe. In the Glasgow area it was recognised for the first time in 1986 by Keith Watson. We still know of only three places: a disused railway at Carmyle, a coal bing at Cardowan and the bings at Hallside Steelworks.

Selfheal (*Prunella vulgaris*)

Geoffrey Grigson tells us that 'The rounded middle tooth of the upper lip of the calyx resembles a hook and was taken as the signature of a vulnerary herb' (a cure for wounds). Discussed in an article in *The Glasgow Naturalist* several years ago by Peter Macpherson, the doctrine of signatures was the divinely inspired belief by physicians that if for instance a particular plant had leaves with the outline of a kidney then that plant should be used to treat kidney trouble. Though the heyday of this doctrine was long ago, there are still practitioners in Glasgow. If they collect their own plants they should have little trouble in gathering Selfheal, a very common plant both locally and nationally. It grows in even the furthest flung parts of Great Britain: the Scilly Isles, St Kilda and Unst. There are records from eighty-three of the Glasgow squares.

Selfheal, an avoider of strongly acid soils, grows in lawns where it can remain prostrate and be easily overlooked. It is found in a diversity of other habitats, though not especially connected with bings. However it grows at Hallside Steelworks where, as the painting shows (see Plate 10), the leaves were not the usual green but strongly tinted reddish. It also grows on the site of the demolished Dalmarnock Generating Station.

Goldenrod (*Solidago virgaurea*)

This is the native small species (see Plate 12) which should not be confused with the much coarser, aggressive invaders, natives of North America: Canadian Goldenrod (*Solidago canadensis*) and Early Goldenrod (*Solidago gigantea*) both very commonly cultivated.

Recorded from only eight squares in the Glasgow area, Goldenrod shows a strange pattern. Seven of the squares are in the south-east, along the banks of the Clyde. The remaining one is at the north-east corner, where Goldenrod grows at the foot of Waterside coal bing. Early last century, Thomas Hopkirk thought Goldenrod was 'rare' and fifty years later Roger Hennedy called it 'frequent' and knew it at Garscube where it has not been recorded recently. Goldenrod is a genetically variable species with a wide tolerance of soil reaction (strongly acidic to alkaline). It is not too easy to account for its unique pattern around Glasgow and to explain its absence from much of the area.

Wild Roses (*Rosa* spp)

Everyone can recognise a Rose but exceedingly few can be very precise about which Wild Rose is which. There is a difficult tangle of Dog Roses (*Rosa canina* group) and Downy Roses (*Rosa tomentosa* group). Some of the differences between one Dog Rose and another or one Downy Rose and another are very subtle: the shape and toothing of the leaves which may or may not have fine hairs and glands, and be aromatic or not. The difficulty is greatly increased by the ease of crossing between and within the groups. Many a Wild Rose is a hybrid.

If the Wild Roses are thought of as one, then it can be said that the Wild Rose is very common in and around Glasgow. There are records from almost all the squares. The habitats include hedgerows, disused railways, scrubby pasture and coal bings as at Waterside and Gateside and the slag heaps at Hallside Steelworks where the illustrated Dog Rose was found (see Plate 9). That it is a Dog Rose, or at any rate has a lot of Dog Rose in its parentage, is clear by the single size of teeth on the leaves and especially by the strongly hooked prickles on the stem. The ripe hips are very important in most precise identifications of Wild Roses.

Helleborine Orchids (*Epipactis* spp)

No less than three Helleborine Orchids grow on two of our coal bings. One is the very common Broad-leaved Helleborine, discussed in detail in the section on gardens. The other two, Young's Helleborine (*Epipactis youngiana*) and Dune Helleborine (*Epipactis leptochila var dunensis*) are of outstanding interest; perhaps they are the most extraordinary discoveries made during the 1980s survey. They exist only in small numbers and nowhere else in Scotland. Searches of wooded bings elsewhere in central and southern Scotland may turn them up.

First found in the Glasgow area in 1985 by Andrew Fullarton, Young's Helleborine (see Figures 15 and 16) was described as new to science only in 1982. A British endemic, it had only a few previously known occurrences in northern England, some of them bings. Dune Helleborine (see Figures 17 and 18), as its name indicates, is sparse and largely coastal in Britain where it is endemic. Its few inland localities are all on bings.

Young's Helleborine may be a stabilised hybrid of very recent origin. For the English populations John Richards' theory is that Broad-leaved Helleborine had crossed with Green-flowered Helleborine, yet another *Epipactis* (*Epipactis phyllanthes*). However, there is no Green-flowered Helleborine in the Glasgow area or indeed elsewhere in Scotland. Perhaps our Young's Helleborine is a hybrid of Broad-leaved and Dune Helleborine. If that speculation is correct then it is not Young's Helleborine at all but a newly evolved species. Its vernacular name could be Bing Helleborine or Glaswegian Helleborine!

On no account should either of these Orchids be disturbed, far less dug up.

Coups or rubbish tips are in the most recent jargon 'landfill sites'. They are the haunt of a very few devotees like Robert Grierson who lived seventy years ago. He hunted the foreign plants that spring up however briefly on coups and found sprouting Tomato, Potato, Date, Orange, Vine and Castor Oil plants and a great many more. He discovered strange-named plants that have not been recorded again in the Glasgow area: Scarlet Horned Poppy and Annual Rock Jasmine (both natives of southern Europe), Amaranth-like Axyris (Russia), Tarweed (Chile) and Small-flowered Dragonhead (Eurasia). The excitement of the hunt for these exotic creatures more than makes up for the unsalubriousness of the edges of the coups! Many coups, having filled the available space, have been grassed over or built upon. In 1916 Robert Grierson searched 'a specially prolific coup between Possilpark and Cowlairs'. By 1931, however, he had fallen foul of the new practice of incineration of the city's rubbish. He moaned, 'The city is burning most of its rubbish now; possibly this is good for the public health, but it is ruinous for my work.' In Grierson's day, long before domestic central heating, coal fires produced huge quantities of ashes which were a major component of the coups. Now coal ashes are a trivial part but there are great amounts of plastics which blow around with very unsightly results at the few operational coups in the Glasgow area: Balmuildy, Wilderness and Cathkin. After decades of incineration, once again the city dumps its domestic rubbish which has first been compressed into large blocks, then made into massive, smooth mounds at Balmuildy and elsewhere outside the area. Up to 1,000 tonnes a day are treated in this way and covered in a round-the-clock operation within fifteen centimetres of brought-in subsoil. Unconsolidated material transported in skips is deposited at Cathkin.

Some components of the coups may change over the years but alien plants still turn up, having sprouted from the remains of food, from thrown-out seed for pet birds, gerbils and hamsters and from outcast seeds and stems of garden plants and weeds, some of which can be aggressive.

The present day hunter *par excellence* on the coups is Peter Macpherson, who has added many plants to Robert Grierson's list. Such a one is an annual that rarely sets seed with us: Probst's Goosefoot, probably originating in America. At Cathkin and Wilderness, he has recorded garden plants like Columbine, Borage, Shasta Daisy, Dotted Loosestrife and Garden Candytuft, as well as pet seed plants such as Safflower, Common Millet and Annual Sunflower. Fly-tipping is unauthorised disposal of rubbish, whether domestic or industrial. Areas of urban dereliction are often used for fly-tipping. Similar plants to those growing on official coups are to be found. Plants such as Canary Grass, Buckwheat, Cereals and Annual Sunflower indicate that remains of pet food is often dumped. In 1989 a small fly-tip at Rigby Street, Carntyne, was the habitat of a single plant of Bristly Oxtongue, a species which is both southern and lime-loving in Britain. It has very seldom been recorded in Scotland. How it reached a small pile of rubbish in Glasgow is a minor mystery.

Yet another kind of dump, the product of Glasgow's industrial past, not being mounded cannot be called a bing nor is it a disorderly pile of rubbish so it is hardly a coup. There is an extensive deposit at Ballochmill Street, produced by the Clyde Paper Works, now closed down. Flooded in winter the whitish spread of paper is only patchily colonised by plants such as Silver Birch, Goat Willow, Soft Rush and Procumbent Pearlwort.

CHAPTER 9

Golf Courses and Heaths

My geographer colleague and keen golfer Robert Price has recently published a book on Scottish golf courses. He has this to say by way of introduction to the courses in Strathclyde Region and Greater Glasgow in particular.

> Greater Glasgow consists of the City of Glasgow and its suburbs and surrounding areas from Dumbarton and Kilbirnie in the west to Kilsyth, Cumbernauld, Airdrie and Wishaw in the east, and from Milngavie in the north to Eaglesham and East Kilbride in the south. In this sub region there are 80 golf courses within a 15-mile radius of the centre of Glasgow (George Square). Within 30 miles or approximately one hour's drive by car from the centre of Glasgow there are about 120 golf courses, which makes Glasgow the leading city in Europe in terms of accessibilty to golf courses.

Perhaps area for area this is the greatest density of golf courses on Earth. Whether that claim can be substantiated or not, the naturalist and conservationist cannot ignore golf courses in west-central Scotland. The 360 square kilometres of the Glasgow area as defined in this book contain thirty-seven courses which cover no less than four and a half per cent of the ground. Classifying the thirty-seven courses by the lie of the land Robert Price lists nineteen on drumlins, ten on 'undulating' ground, four on river terraces, two on raised beaches and two on hillsides. This diversity of landform ensures variety of vegetation. Robert Price places the thirty-seven courses in two broad vegetation types: thirty-four of 'parkland' and three of 'moorland'. Even the very closely mown fairways are not totally devoid of botanical interest but the coarser roughs have much greater appeal to the naturalist. Many a rough is a fragment of heathland. Heathlands are vegetation in which Heather and related dwarf shrubs are dominant and taller shrubs and trees are sparse or absent. Such heathy vegetation often contains a richness of plants, some of which are uncommon or rare.

The botanical interest of golf courses is not only in the heathy ground, which often merges into boggy places. There are often areas of woodland, coniferous or deciduous, and some courses have ponds or lie along riverbanks.

Bitter Vetch (*Lathyrus montanus*)

Bitter Vetch (see Plate 11) grows in only twenty of the ninety Glasgow squares, all but two being in the south. It grows at the edges of woods as at Douglas Park Golf Course and at heathy places as on the rocky knolls at Kirkhill Golf Course. Last century, Roger Hennedy knew it at the Kelvin Woods, Kenmuir banks, near Carmyle and in a wood near the long-disappeared Buttery-burn Loch, High Burnside. From all of these places, Bitter Vetch has vanished a long time ago.

Germander Speedwell (*Veronica chamaedrys*)

In seventy-eight or more of the squares, this is the commonest of the species of Speedwell in the Glasgow area, though Thyme-leaved Speedwell comes a close second. Its habitats are hedgerows, woods and grassland. A plant of waysides, Germander Speedwell (see Plate 11) speeds you well as you travel, as pointed out by Geoffrey Grigson.

Pignut (*Conopodium majus*)

Pignut or Earthnut (see Plate 11) is so called because of the underground, edible tuber. It is a common native plant, growing in sixty-three or more of the squares in such habitats as not very shady woods, parks and grasslands.

Downy Oatgrass (*Avenula pubescens*)

In Britain, Downy Oatgrass is locally common, mainly on limy soils and not found on strongly acid ground. Thomas Hopkirk knew it as rare in pastures, without quoting places. Roger Hennedy knew it at the Cathkin Braes where it may have stayed unseen till 1987 when it was re-found by Graham Steven on the golf course. It also occurs nearby on a basaltic outcrop where it is a very attractive grass when the massed clumps are in silvery bloom with yellow anthers (see Plate 11).

Green-ribbed Sedge (*Carex binervis*)

This Sedge (see Plate 11) has a characteristic look, with its rather stiff fruiting stems held at an angle well off the vertical.

Being very much a plant of heathy and rough, grassy places but not of deep peats, it is a rarity close to Glasgow. It grows only in a few places such as up the Cleddans Burn, near Drumchapel, and the Cathkin Braes. It is no longer found at Stepps Station, beyond Millerston, as reported by Roger Hennedy.

Sheep's-bit (*Jasione montana*)

The geographical range of Sheep's-bit (see Figure 14) is mainly western in Britain but curiously patchy. There are many records from Shetland but hardly any from Orkney and none at all from Caithness, Sutherland and Ross. It grows sparingly in Angus but not in Fife or any part of south-eastern Scotland. Inland in west-central Scotland it is rare.

Two hundred years ago David Ure found it on 'way-side near Galloflat, Hamilton-Farm' (west of Rutherglen). It was in the south-eastern part of the Glasgow area where it was recorded by Roger Hennedy and John Lee at Tollcross; no doubt it grew on the sandy ground, as it probably did at Cadder, also listed by John Lee. It was taken as extinct in our area until a small population was found on small gravel ridges, all but enclosed by westernmost Kirkintilloch. Then, in 1989, Keith Watson found it on low, rocky outcrops just north of Garthamlock and in 1990 he found it at Mount Vernon.

Greater Butterfly Orchid (*Platanthera chlorantha*)

Two Butterfly Orchids have been recorded from the Glasgow area, Lesser and Greater. The Lesser is outwith human memory. It was formerly recorded from Possil Marsh and the Cathkin Hills. Perhaps it will appear again at these places or elsewhere. The Greater is still to be found in eight squares, mostly in the south-west. It inhabits three golf courses (Littlehill, Cathcart and Whitecraigs), heathy ground and grassy places such as a north-facing railway enbankment. In the vicinity of the demolished whisky bond at Cardowan nearly 500 flowering stems were counted in 1988. Near Kennishead, an aberrant form of this Orchid is known (see Figure 19), with deformed flowers. Orchid flowers are considered to be highly evolved for insect pollination. The parts of the flowers are fused together in a complex way and there is only one plane of symmetry, straight through the middle of the flower as one looks into it. Flowers thought to be less highly evolved do not have the parts joined together and there are endless planes of symmetry, like cutting a circular cake diametrically in two. The flowers of the Greater Butterfly Orchid from Kennishead have lost the single plane of symmetry, characteristic of all Orchids. The aberrant arrangement approaches that of the diametrically symmetrical flowers, such as those of a Buttercup. Such abnormalities have caused confusion; some botanists have even considered them to be undescribed hybrids or even species.

Adder's-tongue (*Ophioglossum vulgatum*)

This plant is usually considered to be related to the Ferns though its appearance may not remind most people of that group of plants. Recently a Japanese botanist has claimed that Adder's-tongue and the even more strange-looking Moonwort are not Ferns at all but the only remaining descendants of the *Progymnosperms*, a very ancient group of plants that give rise to the *Gymnosperms* (Conifers, Cycads and others).

In Scotland Adder's-tongue is rare and in the Glasgow area

very rare. William Hooker knew this easily overlooked Fern in the 'Oldlee pasture at Possil', James Balfour recorded it from 'Cathcart Woods', Roger Hennedy added Paisley Canal bank and Cambuslang Glen and John Lee mentioned Torrance. It was reported from Hurlet in the late nineteenth century. There are no recent records from any of these places. The plant seemed extinct in the Glasgow area. However, in 1986 Adder's-tongue was discovered on a bushy embankment of a disused railway near Braehead by Keith Watson and the MSC team (see Figure 21). Soon after the area was redeveloped though not before sods were removed to a garden in Renfrew. In 1987 Keith Watson and the team found Adder's-tongue again, this time on Whitecraigs Golf Course.

CHAPTER 10

Cemeteries and Churchyards

Two very large cemeteries, laid out on hilltops in the second quarter of the nineteenth century, are very prominent features in the city. 'There can be no cemetery in Britain as spectacular as the Glasgow Necropolis.' So in 1974 wrote James Curl who continued in glowing terms about the ornate gravestones. The Necropolis has an area of twelve and a half hectares. Even larger at nineteen hectares is Sighthill Cemetery, again with many monuments to Victorians. Many new cemeteries were necessary by the nineteenth century and into the early twentieth century to serve the needs of the large, increasing populations of British cities. The most recently laid out cemetery The Linn (in 1961) is also the largest in area at twenty-four and a half hectares. In all, there are some thirty cemeteries, large and small, in the Glasgow area. For the most part, the many churches in the Glasgow area, being Victorian or later, do not have graveyards attached. The parish churches of the former villages and even burghs swallowed up by the expanding Glasgow have burial grounds.

Churchyards and cemeteries are not just the resting places of the dead but havens for wildlife, as has been stressed in general terms by Oliver Rackham and Oliver Gilbert and by David Goode for London, where in the absence of maintenance some

Victorian cemeteries have developed into woodland. In Glasgow only St Peter's Cemetery in London Road has in part been overgrown by trees such as Sycamore and a dense growth of Japanese Knotweed.

The seven parish churchyards visited were New Kilpatrick, Cadder, Maryhill Old, Govan Old, Rutherglen, Cambuslang Old and Carmunnock. None supports an especially colourful array of plants, though assiduous mowing by preventing flowering makes the diversity seem less than it really is. However, they do not altogether lack botanical interest. The deconsecrated Maryhill Old alone has Red Campion among the forty species found there and Barren Strawberry, known from only twelve squares, inhabits Cambuslang Old, along with twenty-two other species. Among twenty-nine species, Cadder has Hard Fern and Ramsons. The latter also grows at New Kilpatrick with forty-one other species. Carmunnock is disappointing with only twenty species though it has been invaded by Slender Speedwell, as has New Kilpatrick and Govan Old. There are only some thirty-seven very common species growing in the yard of the last named church.

Botanically by far the most interesting of the large cemeteries is the Necropolis where the lawns are the only habitat in the Glasgow area for Heath Pearlwort which grew formerly at four other places. Even more noteworthy is a Hawkweed (*Hieracium strumosum*) which grows on south-facing sunny slopes; there is only one other Scottish locality of this plant which has a very southern pattern in Britain. Sparse away from the southern Glasgow area, the Fern Polypody grows on a single gravestone. Also found at the Necropolis are Broad-leaved Meadow-grass (only eight squares), Soapwort (only four squares), Black Nightshade (only two squares) and Stag's-horn Clubmoss.

Stag's-horn Clubmoss (*Lycopodium clavatum*)

The discovery of this Clubmoss (see Figure 22), usually a plant of heaths and moorland, no less than five times is one of the surprises of the survey. At the Necropolis it grew on a steep, west-facing bank. It inhabited a lawn at Ruchill Hospital, disused sidings in the Garngad, old ploughed ground at South Pollok Recreation Ground and wasteground at Braehead, near Renfrew. Though it has colonised coal bings elsewhere in Scotland including Bellshill, close to the eastern limit of our area, we have no records from the thirteen coal bings we investigated.

Common Bistort (*Polygonum bistorta*)

The most intriguing botanical puzzle of the large cemeteries is the often profuse occurrence of Common Bistort (see Figure 20), a plant known from forty squares. Two hundred years ago, David Ure reported Common Bistort 'in the east end of Shawfield-bank; in waste ground near Kilbride and in a bank at Castelmilk in great abundance'. In Scotland as a whole it is an uncommon plant which is considered an introduction in some areas such as Angus and Knapdale. In the Glasgow area it has not been found in the yards of the parish churches, nor in the very small cemeteries (all pre 1800) nor in any of the cemeteries laid out in this century except those adjacent to nineteenth-century graves where the plant grows. Moreover it is absent from cemeteries laid out after 1875 except Sandymount (1887) where it is only in small amounts and may have colonised from a population just outside.

Apart from the cathedral grounds where burials took place

till 1898 Common Bistort occurs more or less abundantly in the Necropolis (opened 1832), Sighthill (1840), the Eastern Necropolis (1847), the Southern Necropolis (1840) and St Peter's Cemetery (1857).

Why is there this profusion in the mid nineteenth-century cemeteries? An obvious thought is that it was planted and that may well be the explanation. In northern England, where Common Bistort is commonest in Britain, it has religious and culinary connotations: Easter Ledger Pudding, Passion Dock, Easter Mangiant (see Geoffrey Grigson *The Englishman's Flora*, an engaging book on the vernacular names and folklore of plants). These traditions, however, are not Scottish. In 1831 the need for large cemeteries was stressed by John Strang who wrote a small book, *Necropolis Glasguensis*. On pages 46 and 48 he discussed which plants should be planted at the Necropolis. He does not mention Common Bistort. The large plants of Colchic Ivy on south-facing rocks may be descendants of original plantings. Even more telling is the list of 118 herbaceous perennials given in the book *On The Laying Out, Planting and Management of Cemeteries* published by J C Loudon in 1843. If Common Bistort was a favoured plant for cemeteries why is it absent from this very long list? If it was planted in the large mid nineteenth-century cemeteries in Glasgow who did it and why remains obscure.

CHAPTER 11

Gardens

According to Oliver Gilbert no less than seventy-eight per cent of houses in England and Wales have private gardens. This is a higher proportion than in Belgium, France, Holland and Germany and also more than in Scotland where he stated there are a lot of gardenless tenements and flats. However, all tenements in Glasgow have back greens and often front gardens, if only small. The huge council housing developments built in the 1920s and 1930s in such areas as Mosspark, Knightswood and Carntyne have many thousands of houses with gardens to both back and front.

The recent *City Profile* stated that in 1986 the population of Glasgow District was 725,000 living in an estimated 287,000 households. We do not know the approximate, let alone the exact number of private gardens. However, it must be a very large number which becomes even larger when all the many houses with leafy gardens in the parts of the nine other surrounding districts are added. The urban natural historian cannot ignore private gardens.

Traditionally, Floras have paid little if any attention to the plants of private gardens. Such books have been more concerned with the countryside rather than towns. Into the bargain, it takes an exceptional enthusiast to ask strangers to allow access to make lists of weeds from their gardens. During the Glasgow survey we tried to make such lists. Though the number of gardens studied is only a small fraction of one per cent of all the gardens, the results were rewarding and strongly suggest that many an unexpected wild plant thrives often unnoticed even by the owner, quite apart from any field botanist. Nearly every list we have contains plants of interest.

The diminutive alien New Zealand Willowherb presents a total contrast with all the native British Willowherbs, most obviously Rosebay Willowherb with its patches of tall stems bearing large bright flowers. This invader from the Southern Hemisphere is inconspicuous in its creeping growth, with tiny leaves and dull flowers. After escaping from cultivation by its plumed seeds early this century, it has spread very successfully and often looks totally natural by streamsides and on shady walls. Though it can be found in such places as the terraces of the disused bandstand in Bellahouston Park, in the Glasgow area it is mostly an inoffensive and temporary garden weed. It forms little patches at the edges of paths, in rockeries and such places.

Some noteworthy plants arrive accidentally in packets of seeds or in the soil of container-grown plants. Such is probably the case with the more or less prostrate, tiny-flowered Lesser Swine's-cress, which has a pungent smell if crushed. This Scottish rarity, with very few Glasgow records over the years, was found in a back garden in Mosspark and in a scruffy flowerbed outside a commercial garage in Springboig Road. A much more attractive small plant is Scarlet Pimpernel. In the west of Scotland not often found inland, its Glasgow occurrences are mostly in gardens: in Carntyne, Langside and Milngavie.

There are some curiosities among the wild plants in the gardens in Bearsden and Milngavie District. In 1989 a flowerbed in a terraced front garden in Kessington was the habitat of a single plant of Bristle Club-rush, very small and perhaps often overlooked. It is uncommon in the Glasgow area and normally grows in bare, marshy ground and pastures. Perhaps it has never been recorded from a garden before. Two gardens in Bearsden support Caper Spurge, a plant more at home in southern Europe than in Scotland where it is rare. It sprang up in one of the gardens of a newly built house on a greenfield site. Perhaps the seeds had been dormant in the soil following former cultivation; this agrees with the suggestion by Edward Salisbury that the seeds have a capacity to survive protracted burial. There have been other occurrences of Caper Spurge in gardens at Pollokshields and Jordanhill. It also grew briefly on a fly-tip at Bunhouse Road by the Kelvin Hall, on wasteground now made into a car park. Such plants, even if they survive only temporarily, are part of the wild flora of a city's flora and need to be recognised as such. How many others are waiting to be found by a thorough survey of small private gardens?

Many garden plants, especially herbaceous perennials, can spread and choke out other less aggressive plants. Some of these invade outside areas by seed. They are also often dug up and thrown out by disenchanted gardeners. Several have become some of the commonest, most conspicuous wild plants in the Glasgow area as elsewhere in Britain. Michaelmas Daisies, Perennial Cornflower, Shasta Daisy and Dotted Loosestrife are some of the most successful.

How many residents of the Glasgow area have looked in their gardens for a wild Orchid? How many would even realise that it is possible for there to be such an invader? Nevertheless it is true and there may be hundreds of gardens where the tall, drab, shade-tolerant Orchid, Broad-leaved Helleborine, grows and has grown for many years. Many a plant of this Orchid may well have been consigned to the compost heap or plastic-lined dustbin.

Broad-leaved Helleborine (*Epipactis helleborine*)

The purpose of the kind of botanical recording carried out around Glasgow is far from being just to hunt for rare and threatened plants, engaging though that is, but to assess the frequency of all the plants from the rarest to the commonest. Surprises await the square-bashing recorder who may think that he has already a good idea of how rare or common a particular plant is. Just such a surprise concerned Broad-leaved Helleborine (see Figure 23).

Early last century Thomas Hopkirk knew this Orchid in pastures near his home and he had found it to be 'plentiful' in the Bothwell Woods, where it still occurs today. It thrives there and can sometimes be seen up to one metre tall with 100 flowers; usually it is less than half that height with far fewer flowers. Neither Roger Hennedy nor John Lee knew it nearer Glasgow than Dougalston.

The only hints that Broad-leaved Helleborine might prove common in Glasgow came in the 1960s when there were published reports from Pollok Estate and Maxwell Park. Now we have records from forty-seven of the Glasgow squares. This makes it the second commonest Orchid in Glasgow, exceeded only by Common Spotted Orchid (in no less than seventy-six of the squares, itself a surprise).

Walking a distance of less than 800 metres along the south side of Terregles Avenue in Pollokshields between Maxwell Park and Pollokshields West Station in 1987, Miles O'Reilly found about 200 stems of the Orchid. In 1988 there were only about 100; Orchid populations fluctuate greatly from year to year. Also living in Pollokshields, Brian Knights, one of my colleagues, has recorded the Orchid in several front gardens, while out for an evening stroll. Clearly there are a lot of plants of Broad-leaved Helleborine in that district, only four or five kilometres from the city centre. However, it grows even nearer the heart of the conurbation. John Lyth found two plants growing under a tree in Carlton Place, less than one kilometre from George Square. It grows among shady gravestones at the Necropolis, in the campus of Glasgow University, in the Botanic Gardens, at the fences of the playing fields along Anniesland Road, in Jordanhill College and many other places. Broad-leaved Helleborine does not just have a predilection for small private gardens in Pollokshields. It grows in gardens in Bearsden, Milngavie, Kelvindale, Newlands, Netherlee, Williamwood and Low Blantyre. It is open to being called a snob because all these gardens belong to owner-occupiers. There is not a single record from the gardens of the large council house developments, in particular those to the north-east of the city. We know of it from only one leafy garden in Lenzie and not at all from Kirkintilloch or Bishopbriggs. Map 6 shows the sparseness of the Orchid in the north-east of the Glasgow area. It is a very intriguing scarcity not to be explained in any direct, simple way.

Broad-leaved Helleborine does not just grow in the semi-natural woods as at Bothwell and Dougalston and in some Glasgow gardens. It is an Orchid with a wide tolerance of light conditions and soil, apart from avoiding very acid soils. In 1990 I published a breakdown of its Glasgow area habitats as follows: parks (including cemeteries and golf courses) thirty-five per cent, gardens twenty-eight per cent, woodland sixteen per cent, scrub ten per cent (six per cent on railways, three per cent on bings and one per cent on quarries), three per cent meadows and three per cent roadsides. This is a habitat diversity not indicated in the many field guides to the Orchids of Britain and Europe.

In gardens and parks it is very noticeable that the Orchid often grows where it is not often disturbed: along fences and boundaries, at the edges of lawns and shrub borders. It occurs in rose beds or other places that are only very infrequently dug up. It is a perennial, like all Orchids, and it probably survives below ground even if strimmed away or pulled up. Perhaps it has increased in recent decades because cheap labour is no longer available to keep gardens as tidy as they might be. Moreover its drab greenness gives camouflage.

All Orchids produce large numbers of very tiny seeds that can be freely blown around. In that case, why is the Broad-leaved Helleborine so sparse in the north-east where the dust-like seeds must certainly arrive? That quarter has a lot of arable ground and improved pasture, open water and very acid peat-bogs, all totally unsuitable habitats. Apart from that, if one considers the Orchid to be fundamentally a plant of broad-leaved woodland on the better soils, then that habitat is sparse in the north-east and may have been so for hundreds of years; a late sixteenth-century map discussed in the next chapter shows very little woodland between Glasgow and Kirkintilloch.

There is yet more interest attaching to Broad-leaved Helleborine. It is probably one of the parents of the rare Young's Helleborine discussed in Chapter 8.

PLATE 14
The woods and banks of the Clyde at Bothwell in mid August
× 0.4 natural size

To walk from the Livingstone Memorial Bridge to Uddingston along the banks of the Clyde and through the extensive Bothwell Woods is an experience full of botanical, historical and indeed industrial interest. Even by staying on the path one can see a very rich flora.

On the left of the Plate is a fruiting stem of Hairy St John's-wort. On the right is a stem of Marjoram with only a few flowers left. Marjoram was considered extinct with us because it had not been seen for very many decades. In 1988 it was rediscovered by Keith Watson.

To continue northwards is to enter the Bothwell Castle grounds Site of Special Scientific Interest. However the spectacular ruins of the castle, tidily maintained by Historic Scotland, are not within the reserve. Described by Douglas Simpson as the grandest piece of medieval secular architecture in Scotland and ranking as equal with any contemporary work in England or France, the castle and its surroundings are one of the centres of extinction of plants in the Glasgow area. The ruined walls and towers support the small Fern Maidenhair Spleenwort as well as much commoner plants such as Ivy-leaved Toadflax, Cock's-foot, Dandelions and Rosebay Willowherb. Last century, there were Dittander, Yellow Figwort, Pellitory-of-the-wall and Common Gromwell; all have gone from the castle and indeed the whole of the Glasgow area. The nearby woods once had Spurge Laurel and Butcher's Broom, certainly originally planted, as well as the native Bird's-nest Orchid. The rare Sand Leek still grows on a grassy bank very close to the south-east tower and a hybrid of Perforate and Imperforate St John's-wort, unique in Britain, grows in the meadow east of the castle.

The sloping bank of the Clyde opposite the castle is densely clothed with trees, especially Birch. Only to go there is to realise that the slope is covered with the waste products of coal mining; blocky stones and other debris that even reach the river. Nearby downstream are the all but non-existent remnants of the historic Blantyre Priory where once grew Wild Tulip and Stinking Hellebore. On that slope close to the river you can see Wild Angelica (centre left of the Plate, with white flowers and immature fruits) and with blue-purple, pinkish and white flowers, Giant Bellflower (centre right, bearing nodding young fruits). At the back of the Plate are the fruiting heads of two Grasses, Wood Millet (on the left) and Giant Fescue (on the right). Both have broad leaves characteristic of woodland Grasses. Before reaching Uddingston you may see the only extant Scottish population of Water Chickweed growing with the very similar Wood Stitchwort. Perhaps restricted by climate, the former is very southern in Britain and, though not at all a plant of the hills, the latter is very northern. This may be the only place in Britain where the two grow together.

PLATE 14

PLATE 15

PLATE 16

PLATE 17

PLATE 15
Slender Speedwell
× *1.0 natural size*

PLATE 16
Spanish Bluebell
× *1.0 natural size*

PLATE 17
Primrose
× *1.0 natural size*

Slender Speedwell (*Veronica filiformis*)

An escape from cultivation this century, this is an attractive, creeping little plant with a mass of flowers in spring (see Plate 15). A native of Turkey and the Caucasus, it has been found in thirty-two of the Glasgow squares but may well be in many more, overlooked because it is mainly in private gardens. It can be profuse in large lawns, as at Linn and Rouken Glen Parks, and in small lawns too, as in my back garden where it spreads readily into the herbaceous border.

Its spread both nationally and locally has been achieved virtually entirely without seed which is rarely set in Britain. Lawnmowers make endless cuttings which root easily in our cool, moist climate.

Skunk Cabbage (*Lysichiton americanum*)

A very striking plant for around ponds and streamsides in large gardens, the Skunk Cabbage is a native of North America. Though it emits a smell to attract pollinating insects, it hardly deserves being called skunk and the leaves are not much like those of a cabbage, though they are large and fleshy.

Along the burn that flows through Barloch Moor in Milngavie there are hundreds of plants of Skunk Cabbage which make a splendid display in spring (see Figure 24). It is regenerating very freely by seed, one of which has grown on the east bank of the Allander, one and a half kilometres downstream from the large stand in the middle of Milngavie.

Spanish Bluebell (*Hyacinthoides hispanica*)

This native of the Iberian Peninsula is often cultivated in small gardens and in parks where it meets the native Bluebell. They hybridise freely and the progeny are intermediate, fertile and backcross with the parents. This can easily be seen in gardens in Pollok Country Park and elsewhere. The illustrated plant (see Plate 16) is probably one of the hybrids. Notice the bell-shaped, rather broad flowers, which contrast with the tubular, narrow flowers, all nodding to one side in the native Bluebell.

Dotted Loosestrife (*Lysimachia punctata*)

In 1953 John Lee reported this too vigorous herbaceous perennial (see Figure 25) from 'wasteground near Glasgow'. Now we have records of it from sixty-three of the ninety squares. It has become one of the most commonly encountered garden outcasts; a colourful feature of our wild urban and even rural flora.

Without any details the Sheffield ecologist Oliver Gilbert writes that the dumping of garden rubbish 'most often occurs on wasteground near public houses' and lists several perennials including Dotted Loosestrife. The Glasgow recorders have not been alive to the possibility of such a connection, which needs investigating. Certainly in Glasgow there is no shortage of either Dotted Loosestrife or pubs which sometimes remain alone after surrounding buildings have been demolished. According to a recent report, Dotted Loosestrife seems to be infrequent in Midlothian. Can it be that in Glasgow there is a proportionally greater number of diggers anxious to dispose of their garden rubbish in a thirst-quenching way than there is in the capital city? It may only be that the intensity of recording in Edinburgh has been less than that in Glasgow and so the plant's occurrences have been underestimated. Another possibility is that the cloudier, moister climate of Glasgow makes establishment and spread easier.

CHAPTER 12

Woodlands

In the Glasgow area now there is very little woodland, whether broad-leaved or coniferous, whether modern plantations or wildwood, the vivid term used by Oliver Rackham. Before man the agriculturalist began to clear away the woodlands several thousand years ago, like the rest of Britain the Glasgow area was very largely covered by trees. In our low-lying area there were no serious impediments to the good growth of trees. The predominant trees were Oak, Birches, Wych Elm, Alder, Hazel, tall Willows, perhaps Ash, but very little Scots Pine. Only the very acidic wetness of the peat-bogs and the instability of sand and shingle and perhaps a few cliffy places along the rivers were not part of the great stretches of woodland more than 5,000 years ago. Two thousand years and more ago the Iron Age Celts made substantial though not necessarily permanent clearances and Medieval and later men completed the transformation.

Now only about four per cent of our area is tree-covered. This is less than half the average for Scotland of about ten per cent. Not even a tiny part of the four per cent can be said to be truly wildwood, that is natural woodland unaffected in any way by man's activities. Much is very obviously plantation, and recent plantation at that, with alien trees like Larch, Spruce, Beech and Sycamore. However, to claim for that reason that our woodlands are uninteresting, and not worthy of conservation, would be totally wrong. Even if at best semi-natural their ecological and conservational interest is high. A twentieth-century conifer plantation is less interesting than an old wood of Oak or other native trees; the older the wood the greater the interest.

In an old wood the trees are not necessarily very old. Though some Oaks survive for almost 1,000 years, most Oaks in Britain die in their third century if not before. Not very far from Glasgow, though strictly outside our area, there are two places with very old Oaks. At Cadzow by Chatelherault Country Park there are a few *pollarded* Oaks among the ancient Oaks that were once part of an old landuse called wood-pasture; the tops of the trees have been cut off repeatedly. Most of the old Oaks at Cadzow have naturally lost their tops and from their strange-looking shapes and annual rings their long, chequered histories can be revealed; this is the task of my research student Martin Dougal. The crooked boughs on massive, gnarled and hollow trunks are reminiscent of the Ents in Tolkien's *The Lord of The Rings*. These Oaks are among the most notable in Britain. This is a claim which also applies to the *coppiced* Oaks in Mugdock Wood, part of Mugdock Country Park to the north of Glasgow. Like some of the Oaks at Cadzow they have been cut many times but at soil level. Multiple trunks result and with each cut the new trunks spread further from the original centre. Some of the rings of trunks at Mugdock have spread so far that the trees must have begun to grow hundreds of years ago.

Old maps are treasure troves of information, not least concerning woodlands. In 1795 Thomas Richardson published a map called called *A Map of the Town of Glasgow and Country 7 mile Round*. This is a map that would have been used by Thomas Hopkirk when he was hunting plants in and around Glasgow. It shows plantations as well as semi-natural woods.

The maps made by General Roy to help control Scotland after the '45 rebellion are also very useful, with much detailed information. Even older and therefore of greatly increased interest are the maps in John Blaeu's *Atlas Novus* published in 1654. The Scotsman Timothy Pont, the surveyor for these maps, has with justification very recently been called 'one of Scotland's most unsung heroes'. He had carried out the work by 1596. If one can be sure about pin-pointing a particular wood then that wood is at least 400 years old. Timothy Pont's achievement in covering the whole of Scotland was admirable but such were the methods of surveying and map-making 350–400 years ago that there are severe limitations to what one can reliably deduce from the maps. Garscadden Wood is clearly shown on the Richardson and Roy maps and much less certainly on the Pont map. With much more confidence Cadder Wood (tree symbols at 'Caldar Caft') is on the Pont map and so are Mugdock ('Mugdak'), Crookston ('Krukstoun wood'), Pollok, Langside ('Langsyid'), Linn Park and Bull Woods and some others.

Some woods on the old maps have completely disappeared. Examples are Rutherglen Farm ('Ruglan Farm' on Pont's map) and 'Earlston Wood' between Paisley and Glasgow (on John Watt's map of 1734).

For a wood to be firmly identifiable on an old map does not mean that the wood has stayed unchanged at the hands of man since the time of the map. Usually this is far from being the case. A wood may have been clear-felled and replanted with different types of trees. Crookston Wood is now a recent coniferous plantation. The present Bull Wood is divided in two by a boundary wall. To the west it is in part planted coppice Oak over runrig, the old strip arable cultivation method and to the east there are planted conifers.

Trees, interesting as they are, are far from being the only plants of note in a wood. The smaller inhabitants have great attraction for the natural historian and conservationist as the examples discussed and illustrated show. Garscadden and Bothwell Woods have been singled out. However, all the woods are notable in one way or another.

Primrose (*Primula vulgaris*)

Often grown in gardens, Primrose is a very familiar native plant (see Plate 17). Around Glasgow it grows on grassy banks, riversides and in woods; nowhere can it be said to be abundant. In heavy shade Primrose does not flourish nor can it regenerate well by seed on very acid soils.

The pattern of Primrose around Glasgow is very striking. All the squares in which it grows are round the margin and mostly in the south. The three populations on the north are small, more often inconspicuous and difficult to reach. These remarks apply to many of the southern stands also.

Gardeners often dig up Primroses from the wild. The rangers at Chatelherault Country Park, near Hamilton, recently caught a married couple carrying off a wheelbarrow-load. Such activities are blamed for scarcities around other cities such as London. However, the well-known woodland ecologist, Oliver Rackham, questions this assumption and thinks that change in woodland management has been very important. The former sustainable exploitation, coppicing, produced unshaded periods long enough for Primroses to thrive. Now our woods are too continuously shady.

Lesser Celandine
(*Ranunculus ficaria*)

Usually on fertile soils, this spring-flowering, low-growing plant (see Plate 18) inhabits riverbanks and woodlands and often extends on to nearby meadows. It grows on the very steep, west-facing bank below Bothwell Castle. It is inherently very variable. Often growing in shady places, some plants produce bulbils at the base of the leaf stalks. They are genetically different from those plants without bulbils as in the Plate.

There are records of Lesser Celandine from forty-five of the Glasgow squares, more than two-thirds being in the western half and less than one-third in the eastern half. By the end of June at the latest, all the above ground parts have withered away and only the tuberous roots remain, ready to grow again the next spring. Its presence can therefore go unseen by recorders during summer.

Keeled Garlic
(*Allium carinatum*)

Onions, Leeks and Garlics, including the culinary one, are all in the genus *Allium*. Some species of *Allium* have attractive flowers and are grown for that reason, as is the case with Keeled Garlic, which can have a mixture of flowers, that rarely produce seeds, and bulbils in each flower-head (see Plate 19). It is a native of much of Europe but not of Britain where it has escaped from gardens, although not very often. The wild occurrences of

Keeled Garlic in Scotland are mostly coastal as in Ayrshire.

Only since August 1989 have we known it wild in the Glasgow area. Well away from any gardens, there is a small patch by the edge of the walkway through the Bothwell Woods, on the edge of the Clyde. Though the learned *European Garden Flora* warns that Keeled Garlic can become a weed because of the bulbils, its discovery by John Lyth was a fitting end to the last official excursion for the Flora of Glasgow Survey.

Toothwort
(*Lathraea squamaria*)

In the British flora there are only a few fully parasitic flowering plants. These are completely devoid of the green pigment chlorophyll and cannot make their own food by photosynthesis. There are ten Broomrapes (*Orobanche* spp) in Britain but only three in Scotland and none within the Glasgow area. There are a few species of Dodder (*Cuscuta* spp) very rarely encountered in Scotland. Flax Dodder, as its name indicates, was a parasite of Flax crops. Last century it was known at Chryston. The crop has vanished and so inevitably did the parasite.

Toothwort (see Figure 26) is the only native parasite that can be regularly seen in the Glasgow area. Last century it grew at Carmyle and Langside but there are no modern confirmations. This century it has been reported from Cambuslang, the banks of the Cart and Pollok Country Park. In Linn Park, where it grows from the roots of Wych Elm, Toothwort is at its only Glasgow locality, where it has been often seen in the 1980s.

Toothwort has a very unattractive dingy white or pink colour. Purple Toothwort, which has been grown at the Botanic Gardens for several decades, is more vivid, as its name implies.

Common Twayblade (*Listera ovata*)

Twayblade means 'two leaves', all each shoot ever has. Its overall greenness makes it an inconspicuous Orchid (see Figure 28). There is a considerable diversity of sunny to shady habitats including woodland as at Darnley Glen, railway embankments as at Lenzie and Kennishead Road, disused railway tracks as at Braehead and wooded bings as at the Milngavie lime dumps.

There are no 1980s records confirming old discoveries from Blantyre Priory, Bothwell, Dougalston or St Germains Loch. However, it may be lurking at any or all of these places.

Sandleek (*Allium scorodoprasum*)

In Britain the alien Sandleek is a northern rarity. A native of mainland Europe and south-west Asia, it has been known in the vicinity of Bothwell Castle for more than 100 years (see Figure 27). It still grows there on a grassy bank close to the south-eastern tower and sparingly on the banks of the Clyde south of the castle. At its only other Glasgow locality, Springhill Farm, north of Garrowhill, it is only in small numbers by a streamside.

Hairy St John's-wort (*Hypericum hirsutum*)

In Britain Hairy St John's-wort (see Plate 14) is common in the south where it is strongly connected with limy soils but uncommon and eastern in Scotland. In the Glasgow area it is in thirteen squares: ten to the south and east of the city and three at and near Renfrew. The habitats in the south-east are mostly the banks of the Clyde, where David Ure saw it 200 years ago. At Braehead, near Renfrew, it has colonised a disused railway.

Marjoram (*Origanum vulgare*)

Like Hairy St John's-wort, Marjoram (see Plate 14) is a southern lime-lover in Britain and in Scotland it is uncommon. In the Glasgow area only one small clump is known. In west-central Scotland it may be limited both by the wet, cool climate and the prevalence of strongly acidic soils.

Wild Angelica (*Angelica sylvestris*)

In the Glasgow area Wild Angelica (see Plate 14) is known from seventy-one of the squares, mostly round the margin. Its habitats are mostly marshes and watersides but sometimes it occurs in drier places.

Giant Bellflower (*Campanula latifolia*)

Two hundred years ago David Ure recorded Giant Bellflower (see Plate 14) on 'Banks at Calderwood, and in a hedge between Hamilton-Farm & Clyde'. This very handsome Bellflower reaching well over one metre tall has flowers up to five and a half centimetres long. It is known around Glasgow from fifteen squares, all but two in the south. More or less shady banks of the Clyde, Cart and Kelvin are the main habitats. In the north at Garscube and Torrance there are only very small populations, as is true of Kenmuir Wood also.

Wood Millet (*Milium effusum*)

In the woods around Bothwell Castle there is a profusion of this tall Grass (see Plate 14) which is known from only ten of the Glasgow squares, all but one in the south. In Britain as a whole it is locally common, especially in the south. In some areas, such as Sheffield, it grows mainly on strongly acidic soils and in others such as Devon, mainly on limy soils. In Scotland it is uncommon.

Last century Wood Millet grew near the Aqueduct Bridge, Maryhill and the nearby Gairbraid Glen. It is not there now. However, it still grows in Langside Wood where Roger Kennedy knew it in 1865.

Giant Fescue (*Festuca gigantea*)

Growing mainly round the margins of the Glasgow area, Giant Fescue (see Plate 14) has been recorded from half the ninety squares. It is similar in stature and overall appearance to Hairy Brome and the two often grow together; Giant Fescue has hairless leaves and leaf sheaths while those of Hairy Brome are conspicuously hairy.

Giant Fescue is much less common in Scotland than in England. Its favoured habitats in the Glasgow area are woodlands and riverbanks. We do not know it in wetlands, as reported from the Sheffield area 'frequently'.

Wych Elm (*Ulmus glabra*)

This is the only Elm (see Plate 13) that everyone accepts as undoubtedly native in Britain. Certainly this is so for Scotland; the other Elms such as Small-leaved and English are always obviously planted. Fossil pollen shows Wych Elm to have been present for 8,000 years or more. The present habitats are riverbanks, hedgerows, woods and roadsides. Disused railways, coal bings and other wasteground have been colonised by seed but Wych Elm does not sucker like the other Elms. Glasgow parks and avenues have many planted Wheatley Elms. However English Elm has been much less often grown; a narrow plantation at Saughs Road, Robroyston has produced thickets of suckers. The classification and history of Elms is a fascinating but controversial subject. Oliver Rackham's account is both sound and readable.

Pink Purslane
(*Montia sibirica*)

This attractive shade-tolerant plant (see Plate 13) is a native of eastern Siberia and Pacific North America. Although well known in west-central Scotland by the final quarter of last century, it is not clear when Pink Purslane first grew wild in the Glasgow area. By 1933, however John Lee wrote: 'fully established and spreading; abundant in many localities'. Now we know it to be in at least sixty-two of the Glasgow squares, even at the heart of the city. Its habitats are mainly woods and gardens.

Opposite-leaved Golden Saxifrage
(*Chrysosplenium oppositifolium*)

This small, patch-forming plant (see Plate 13) is a *country cousin*. It grows in forty of the Glasgow squares, overwhelmingly round the margin and mostly in the south; it is rural rather than urban. The habitats of this native are often shady, marshy ground and streamsides.

Bluebell (*Hyacinthoides non-scripta*)

As a young Glaswegian I learned to call *Hyacinthoides non-scripta* 'Bluebell' (see Plate 16) and many a wood in and furth of Glasgow was and is 'Bluebell Wood' to the locals. We did not know that lots of books written over many years, perhaps by mainly English authors, assured the reader that in Scotland the Bluebell is *Campanula rotundifolia* (Harebell). Bloody Man's Fingers, Blue Goggles, Crawtaes, Cuckoo's Stockings, Granfer-griggles and Single Gussies are just some of the names for Bluebell compiled by Geoffrey Grigson, who also listed the names in the Celtic tongues. Whatever the name, massed Bluebells are one of the glories of springtime British vegetation. Bluebells are more at home in the mild climate of Britain than elsewhere in adjacent western Europe. The acid soils of Oak woods as at Garscadden and Mugdock Woods and elsewhere support huge carpets. Bluebell also thrives under Bracken, as on the eastern slopes of Ailsa Craig, where no shade is cast by any tree other than exceedingly sparse and low-growing Elder and Aspen.

In the Glasgow area Bluebell is in danger of being badly adulterated by Spanish Bluebell. See Chapter 11.

Scaly Male-fern (*Dryopteris affinis*)

Broad-leaved Buckler-fern (*Dryopteris dilatata*)

Both these Ferns (see Plate 13) are readily found in the Glasgow area and are by no means confined to woods, though in shady woods there can be large stands of Broad-leaved Buckler-fern as in the coniferous plantation on the cut and drained bog, Low Moss, by Bishopbriggs. Both can colonise shady walls and wasteground in urban situations, as at Hyndland Station and High Street in the case of Broad-leaved Buckler-fern.

PLATE 18

Lesser Celandine

× *0.7 natural size*

PLATE 18

Figure 19 *Greater Butterfly Orchid at Kennishead Road.*

Figure 20 *Common Bistort at St Peter's Cemetery. (Photograph J H D).*

Figure 21 *Adder's-tongue at Braehead, Renfrew.*

Figure 22 *Stag's-horn Clubmoss at the Necropolis.*

Figure 23 *Broad-leaved Helleborine.*

Figure 24 *Skunk Cabbage at Barloch Moor, Milngarvie.*

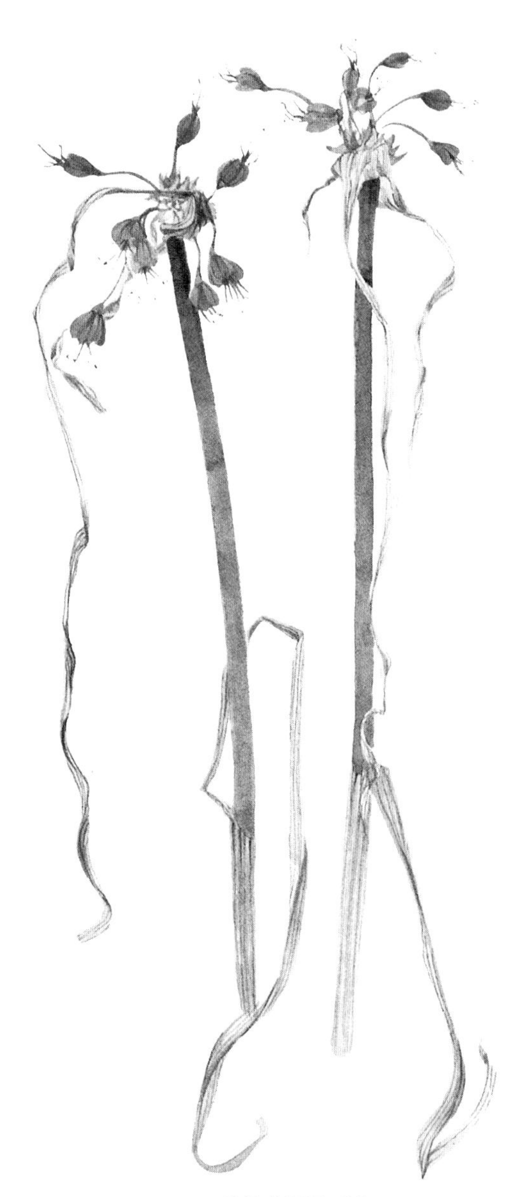

PLATE 19

PLATE 19
Keeled Garlic
× *0.7 natural size*

CHAPTER 13

Wetlands: Fens and Bogs

Especially but not only in the north and west of Britain, our climate, with the large number of rainy days, the often overcast skies and coolness, is very conducive to waterlogging of soils and the growth of peat. It is very difficult for Britons living in the late twentieth century and used to swift travel in dry comfort to envisage just how wet and peaty our landscape used to be, even only a few hundred years ago. Small-scale attempts at drainage and hand-cutting of peat for fuel have gone on for centuries and even millennia, with accumulative effects. Large-scale drainage schemes and deliberate removal of large peat-bogs fall within the last few centuries.

Between Aberfoyle and Stirling, the broad clayey carse lands still support large and deep peat-bogs. Though the natural margins have been cut away long ago, Flanders Moss is still impressive in its five kilometres diameter. A little to the east there was an equally large if not larger bog, Blairdrummond Moss. But now all that remains is an angular-outlined narrow strip, planted up with conifers. What happened was that, from the middle of the eighteenth century, at least 800 hectares were completely cleared away right down through several metres of peat to the sticky clay. This is a particularly well-documented and dramatic example.

There were large raised bogs on the very low-lying, ill-drained land just west of Glasgow, in the vicinity of the airport. Of Linwood, Barochan and Paisley Mosses very little is left that is remotely natural. Within the Glasgow area, peat-bogs, just as much as elsewhere, have been cut away completely or greatly reduced, drained, burned, cut though by railways, planted with conifers and peat extracted on commercial scales. Cutting for fuel began at Lenzie Moss at least as early as 1226.

In the late eighteenth century the parish of Govan still had forty hectares of peat fit to be dug for fuel. This has long since disappeared though place names like Mosspark, Moss Road and Bogmoor Road preserve the memory. As is the case for woodlands, old maps are useful in tracing the former locations and sizes of peat-bogs. The Thomas Richardson map of 1795 shows many mosses near Glasgow, including one between what is now Mosspark and Pollok Country Park and another in the Maxwell Park area. At what is now the Arkleston area, between Glasgow and Paisley, the map locates a triangle of farms called Bogside, Mossland and Honeybog; there is no peat there now. However, to the north-east of the city the map shows several large mosses, all unnamed except Mountain Moss, now called Lenzie Moss. The origin of the name Mountain Moss is obscure, though there is a fine view of the Campsie Fells. All of these north-eastern mosses on the Richardson map have been more or less damaged, some very greatly so, especially Cardowan Moss, the moss between Millerston and Robroyston, Low and High Mosses at what is now Bishopbriggs, and the large bog between Garnkirk and Gartloch. The map also shows lochs at Robroyston and Lochgrog; both were drained long ago.

The wetlands mentioned so far are all bogs in the strict ecological sense. These are wetlands where the few plants tolerant of waterlogged, very infertile, highly acidic conditions are accumulating as dead remains and forming more or less decomposed peat, with little if any inwashed mineral material. Bogs may grow for thousands of years and become slightly domed, the centre being slightly higher than the margins. They are called raised bogs: Flanders and Blairdrummond are examples and so are Lenzie, Commonhead and Garnkirk Mosses in the Glasgow area. Such bogs get water and dissolved nutrients only from the rainfall. Beginning to grow in badly drained hollows, fens can be much less acidic than bogs and much more fertile because of inwashed clay or silt. There are extensive fens at Possil Marsh and Bishop Loch. By natural succession fens can turn into bogs.

Ecologists spend much time investigating all the many sorts of fens and bogs and how they merge both in space and time. Just how a bog becomes domed is a puzzle and it is very difficult to account for the complex, changing patterns of pools and hummocks found on the surfaces of undamaged bogs. Everyone agrees that Bogmosses, with their water-holding and acidifying properties, are crucial in bog ecology.

A column of peat taken from a bog or a column of mud taken from a loch bed is a coded message written over thousands of years. The remains of seeds, leaves, wood, beetles and microscopic pollen grains and cysts and eggs of invertebrate animals can be extracted, identified and counted. From such information, deductions can be made about vegetational and climatic changes and about man's activities in cutting down the forests and growing crops. It so happens that, though peat-bogs in Britain have been studied in these ways at many places, little has been done to investigate the bogs in the Glasgow area. We know a little about Lenzie Moss, where the five metres of peat began to form in the southern part not less than 6,500 years ago. At first there was a damp woodland of Birch, Alder, Oak, Hazel and Willow which gave way to bog with Heather, Hare's-tail, Cotton-grass and Bogmosses. A little below the two metre depth where the peat is unlikely to be less than 2,000 years old, there are clear signs of human disturbance of the area (high values of pollen from grasses and weeds).

Round-leaved Sundew (*Drosera rotundifolia*)

The Sundews are carnivorous plants that respond to trapped insects and tiny pieces of meat or cheese but not for long to fragments of wood or metal. There are three species in Britain, at once fascinating in their unusual feeding habits and very disappointing in their small stature. Despite what can be seen on television or at the cinema there are no man-eating plants known to science. Plate 22 shows a plant of Round-leaved Sundew at its diminutive natural size.

Peat-bogs are the main habitat of Round-leaved Sundew and because such boggy land has been destroyed, many lowland areas have lost this plant. Consequently it is often and rightly quoted as a declining plant in Britain as a whole. In the Glasgow area it has disappeared from Possil Marsh and the Cathkin Braes. The only slightly larger Oblong-leaved Sundew was seen by Thomas Hopkirk in a marsh 'between Glasgow and Paisley'. Both the unnamed marsh and its special plant must have disappeared a long time ago.

Donald Patton and William Rennie were both long-lived local botanists who visited Possil Marsh many times. In a joint publication, William Rennie regretted the activities of Bogmoss gatherers. Bogmoss was valuable formerly as a wound dressing and was still in use as such in the Second World War. It remains important in horticulture for growing tropical Orchids and for lining hanging baskets. As he watched the decline of the Round-leaved Sundew at Possil Marsh William Rennie tried to prevent its demise by introducing plants from other places. He wrote:

> Hennedy, 1865–1890 records it. . . . In 1910 I first seriously noted it becoming fewer and fewer. . . . In 1918 about two score were planted. . . . In 1919 a magnificent show but unfortunately Sphagnum moss collectors removed many. . . . In 1920 less than a dozen plants. In 1925 the stock was renewed and again they gradually disappeared. In 1929 a new lot was introduced and again in 1930 there was a gorgeous display. Moss collectors again appeared. No sundew was seen in 1930. One plant was seen in 1931 . . . the last.

This description of the decline to extinction of a particular plant at a particular place is the only one we have from the Glasgow area. William Rennie would be pleased to know that the Round-leaved Sundew still survives at Cardowan, Commonhead and Lenzie Mosses. He might well be surprised to learn that it has increased its hold on the Glasgow area by invading two stretches of abandoned railways. In the south it grows sparingly in that habitat, while close to the city centre there is an impressively large population in a small area of disused sidings. The partly flooded ballast and sleepers support not just the Sundew but other bog plants such as Bogmosses, Sedges and both Cottongrasses.

Cranberry (*Vaccinium oxycoccus*)

With small, somewhat heart-shaped leaves and thread-like creeping stems, Cranberry is the plant with berries made into a sauce for eating with the Christmas turkey. The tiny flowers are rather reminiscent of the much larger flowers of a Turk's-Cap Lily and give rise to the berries, sometimes speckled but usually not. Whatever their colour the berries always seem ridiculously large, being much larger than the flowers and the leaves. Like Round-leaved Sundew, Cranberry is a bog plant that often grows on top of Bogmosses. It grows at Lenzie Moss, Bishopbriggs (both High and Low Mosses) Possil Marsh and Cathkin Braes Golf Course (see Plate 21). The last named is the only place in the south of the Glasgow area.

Hare's-tail Cottongrass (*Eriophorum vaginatum*)

Common Cottongrass (*Eriophorum angustifolium*)

These closely related plants, being very tolerant of wet infertility, often grow nearby on bogs where the peat may be very deep. Hare's-tail Cottongrass (see Plate 11) forms tussocky growths and prefers the drier hummocks while the extensively creeping Common Cottongrass (see Plate 20) prefers the wetter hollows. Like Bogmosses and Heather, Hare's-tail Cottongrass is one of the major components of peat in which the fibrous remains of its tussocks are preserved for thousands of years and are so tough as to resist the peat-cutter's spade.

The plumed fruits of both Cottongrasses are dispersed far and wide by the wind and begin to grow in places that are not at all peaty. Both have colonised flooded disused sidings near the city centre. In 1984 one tussock was found in the middle of

Auchenshuggle Wood and in 1989 there was a small tussock on the west-facing steep slope overlooking Speirs Wharf in Port Dundas.

Marsh Cinquefoil
(*Potentilla palustris*)

Locally common in the north of Britain, Marsh Cinquefoil (see Plate 20) grows in twenty of the Glasgow squares, all but five in the north. Each leaf has five leaflets; hence the 'Cinquefoil'. The habitats are loch shores and marshes, as at Possil Marsh where the plant is abundant. Its long established occurrences in the Glasgow area are proved by the discoveries of fruits that grew at the end of the last Ice Age at Garscadden some 10,000–11,000 years ago and at Bearsden in Roman times.

Greater Spearwort (*Ranunculus lingua*)

This is the tallest, largest-flowered Buttercup in the British flora (see Plate 20). It can reach well over one metre tall with flowers up to five centimetres in diameter. Plentiful at Possil Marsh where it has been known for nearly 150 years, Greater Spearwort is a Scottish rarity and is uncommon in Britain as a whole.

Mare's-tail (*Hippuris vulgaris*)

Mare's-tail (see Plate 20) is locally common in Britain. Like Marsh Cinquefoil, Mare's-tail is known to have grown at Garscadden at the end of the last Ice Age and it is a plant of marshes and loch margins. It grows in eighteen squares mostly in the north.

Water Horsetail (*Equisetum fluviatile*)

The small group of Horsetails have many distinctive features in their shape, chemical make-up, reproduction and very long history. There are no more than twenty-nine or so species in the world today and the largest reaches up to thirteen metres high but it is a weak-stemmed, sprawling plant. As self-supporting plants they do not now stand very tall but once they were much taller and important parts of the vegetation of the world. That 'once' was a very long time ago, 300 and more million years. You can see remains of these ancient Horsetails in Kelvingrove Museum and in the Hunterian Museum of the University of Glasgow. To visit the Fossil Grove in Victoria Park is to see the petrified remains of a wetland in which the long extinct Horsetails and Giant Clubmosses grew; these great coal forests have been well discussed, with illustrations, by Barry Thomas.

Ten of the twenty-nine present day Horsetails grow in Europe, nine of the ten grow in Britain and eight of the nine grow or grew in the Glasgow area. Water Horsetail, one of the commoner ones, is known from thirty-six, mainly northern, squares. Often it lacks branches altogether; sometimes as shown in Plate 20 it has sparse and straggly ones. It can grow into large stands as at Possil Marsh and nearby fens and at the edges of lochs. At Summerston and Robroyston, it has colonised flooded railways.

Christopher Page has listed the practical uses of Horsetail. He began: 'Certainly by the Middle Ages and probably from Roman times onwards, if not earlier, the practical uses to which Horsetails have been put were many and varied.' He mentioned their being eaten as vegetables, their use as sandpaper for wood, burnishing armour, scouring pots, staunching blood and as an early form of gas mantle. Some of these uses depend on the minute roughness of the surfaces of Horsetails, toughened by the deposition of silica. The Horsetail called Dutch Rush, a national rarity, was especially good as a scourer and gas mantle. For these uses, it may have been over-collected from the banks of the Clyde where it grew 'between Carmyle ford and the mill, plentifully' according to Thomas Hopkirk. Now all that is left are a few stems at the eastern extremity of Kenmuir Wood; in 1989, unauthorised gravel extraction missed obliterating this last, tiny stand by about ten metres.

PART 3

Conservation of City and Countryside Plants

CHAPTER 14

Plants of the Glasgow Area

Orchids have an appeal that few other plants possess. They are considered by many people to be exotic and expensive, brought back from faraway tropical countries such as New Guinea or Brazil for display in steamy hot glasshouses at garden festivals and botanic gardens. It surprises many to learn that there are Orchids native in Britain and indeed other countries with temperate climates. The fifty wild British Orchids are a small number compared to the richnesses in the tropics where a single large tree may support many different Orchids. Nevertheless, even within our limited number the diversity of flowers from the beautiful to the bizarre, the complicated pollination mechanisms, the protracted growth from the dust-like seeds and many other features of Orchid biology fascinate botanists and conservationists. For all these reasons it is easy for conservationists to fight for the protection of Orchids, some of which are rare and liable to be dug up by unscrupulous people. Michael Hutchings has outlined the status of rare and declining British Orchids on a national scale and discussed reasons for the decreases. No less than fifteen of the fifty native British Orchids grow or have grown in the Glasgow area. In national terms some are common, some are uncommon and two are very rare. They are listed in Table 3.

Table 3. Orchids of the Glasgow Area

	British Status	Glasgow Squares
Common Spotted Orchid	Common	75
Broad-leaved Helleborine	Locally common	48
Northern Marsh Orchid	Common in north	32
Common Twayblade	Common	9
Greater Butterfly Orchid	Locally common	8
Heath Spotted Orchid	Common	4
Young's Helleborine	Very rare*	2
Dune Helleborine	Very rare*	2
Early Purple Orchid	Common	1
Bird's-nest Orchid	Uncommon	†Extinct
Early Marsh Orchid	Uncommon	‡Extinct
Fragrant Orchid	Locally common	†Extinct
Frog Orchid	Uncommon	†Extinct
Lesser Butterfly Orchid	Locally common	‡Extinct
Small White Orchid	Uncommon	†Extinct

* Specially protected, found during the 1980s survey for the first time.
† Not recorded since the 19th century.
‡ Not recorded since the 1950s.

More than one-third of the Glasgow area Orchids have become extinct in less than 200 years. Four have not been seen since last century and two since the 1950s. The concern expressed by conservationists about Orchids is well justified by such a large proportion of locally vanished species. However, brief consideration of Table 3 raises the question 'Are all the Orchids in the Glasgow area equally deserving of protection?'

It is illegal in Great Britain for anyone to uproot *any* wild plant without permission from the landowner or occupier. (This does not prevent the picking of Daisies or the gathering of wild

Rasps.) Young's Helleborine is further protected by being listed in Schedule 8 of the Wildlife and Countryside Act of 1981. It is illegal intentionally to pick, uproot, destroy or sell any of such proscribed plants or to collect or sell their seed. The populations of the Dune Helleborine in the Glasgow area are so small that this Orchid is a candidate for the description as the rarest plant in Scotland. These two Orchids are of national, even international, significance.

Early Purple Orchid is another good case to contemplate, but not because it is rare nationally. On reading Floras and inspecting distribution maps one realises that it is common on a national scale. However this Orchid has only one stand in the Glasgow area where it has probably been present for nearly 150 years. Formerly it grew 'below the Cathkin Hills'. In local terms Early Purple Orchid certainly deserves conservation. Three of the Orchids in Table 3 are far from rare, both locally and nationally. Common Spotted Orchid has been recorded in no

---

PLATE 20

Possil Marsh, the southern fen in early July

× 0.4 natural size

There is a partly junk-filled canal and recent housing to the east side, a busy road and large cemeteries to the west side, urban sprawl immediately to the south and a prominent line of electricity pylons. This does not sound much like an attractive place in which to enjoy nature. However, such are the surroundings of Possil Marsh, the only nature reserve in Glasgow District. It fully deserves its protected status, being the habitat of many waterbirds and wetland plants.

For 175 years and more, Possil Marsh, with its suite of rarities, has

been a favourite place of botanists, both amateur and professional alike, including generations of members of the Glasgow Natural History Society. In his distinguished book, Thomas Hopkirk listed many of the famous plants, some of which have long since vanished. At least six species in what was their only locality near Glasgow have become extinct.

The southern end of the marsh is an extensive fen with large stands of Great Reedmace, Bulrush, Bogbean, Bottle Sedge and Water Sedge. In midsummer the fen is colourful with the large Buttercup called Greater Spearwort (yellow flowers), Marsh Cinquefoil (burgundy flowers) and Common Cottongrass (white fruiting heads). With very inconspicuous flowers, Mare's-tail (left of centre) somewhat resembles Water Horsetail (left); this entirely superficial similarity, as well as their common names, leads to endless confusion.

If you leave the canal towpath you will have to struggle through quaking head-high growths of the marsh plants or weave through the scrubby Sallows and Bay Willows at the drier northern end. You might be forgiven for thinking that you were in an unchanging, pristine wilderness, even though it is hemmed in by so many human structures. You would be wrong. Had you walked along the towpath in the mid 1800s you would have seen no less than four ironstone mines within what is now the reserve. The Victorian Ordnance Survey maps mark them as well as several others to the immediate north and across the canal to the north-east at Laigh Kenmuir.

For ornithological reasons, Possil Marsh became a privately owned nature reserve in 1930 and a Site of Special Scientific Interest in 1954. Now it belongs to the Scottish Wildlife Trust (since 1982).

PLATE 20

Figure 25 *Dotted Loosestrife.*

Figure 26 *Toothwort.*

Figure 27 *Sandleek (purple heads, right) at Bothwell Castle.*

Figure 28 *Common Twayblade.*

Figure 29 *Springburn from the southwest in 1988, with the mid-nineteenth century Sighthill Cemetery, bottom left. Reproduced by permission of City of Glasgow Planning Department.*

PLATE 21

PLATE 22

PLATE 21

Cranberry

× *1.0 natural size*

PLATE 22

Round-leaved Sundew

× *1.0 natural size*

less than seventy-six squares; it may be in all ninety. Unless the circumstances are very unusual, an occurrence of this Orchid need not cause much concern on the part of conservationists. What might such special circumstances be? If the Orchid is growing in large numbers, making a colourful display in company with other plants as an unusual assemblage on a bing or disused railway, then the case for conservation may be a strong one.

Why have no less than six of the Glasgow area Orchids become extinct? It is well known in conservation circles that the Lady's-slipper Orchid, a lime-loving species confined in Britain to northern England, has been reduced to a single plant, guarded during the flowering season. The *Red Data Book* states: 'Its virtual extinction has been due to uprooting and picking by gardeners, botanists and others from very early times.' Not only very rare Orchids are dug up. In 1987 Nick Stewart reported that: 'In July over 100 Early Purple Orchids, Green-winged Orchids and Greater Butterfly Orchids were removed from four sites in Gloucestershire. Then specimens of Lady Orchid and Early Spider Orchid were removed from two sites in Kent. All these thefts have the hallmark of someone who knows what he was doing, but no one has been caught, and we do not even know whether the thief was the same in each case. Every year there are reports of specially protected plants being stolen . . .'

Such activities are to be deplored. However, they do not explain the disappearance of the six Glasgow area Orchids. There is no need to envisage that conscienceless collectors or gardeners have been at work. That, however, is not to condone the digging up of Orchids. With the possible exception of the weedy Broad-leaved Helleborine, no Orchid from the commonest to the rarest should be dug up, unless to stop certain destruction by unpreventable developments.

There is no way of knowing with very great precision where the six extinct Orchids grew nor are there any accounts of exactly what happened to extirpate them. However it is beyond reasonable doubt that they vanished because of changing landuse: ploughing and draining the land for farming and forestry, house and road building or even leisure activities by park-making. Three of the Orchids disappeared from the Castlemilk-Cathkin district which in the last 100 years has become more and more urban and has seen the laying out of golf courses and Cathkin Braes Park, acquired by Glasgow in 1886. Two of the six disappeared from Possil Marsh for no immediately obvious reason. A plausible speculation is that, as the landuse has changed, the north end has become less heathy and more scrubby with the Orchids being shaded out. The Castlemilk-Cathkin district has been one of the centres of extinction of plants since last century. Others have been Possil Marsh, the Tollcross-Daldowie district and the banks of the Clyde from Carmyle to Blantyre with the Bothwell Woods and Bothwell Castle itself.

The sandy and gravelly ground between Tollcross and Daldowie was formerly arable fields and unimproved grassland. Many well-known but now much reduced or extinct weeds grew there: examples are Corn Gromwell, Pheasant's Eye, Prickly Poppy, Dwarf Spurge and Field Scabious. On a national scale their declines relate to better seed cleaning and other altered practices and finally to weedkillers in the last half century. On the local scale there were the additional causes of the spreading of the city south-eastwards and the extraction of sand and gravel necessary for building. None of the exterminations at the centres of extinction or elsewhere, leading to complete disappearance from the Glasgow area, can be directly and unequivocally related to pollution whether of the air or of water. In and around Glasgow and other major industrial cities the sulphur dioxide-laden air much reduced the Lichens, well known for their sensitivity, but not the Flowering Plants and Ferns discussed in this book.

In all there are about 100 species including the six Orchids recorded by earlier botanists but not found during the 1980s survey. The major causes of the extinctions were changing landuse: the spread of industry and housing with the making of canals, railways and roads and the modernising of agriculture. A full list of the extinct species can be found in a report I wrote with Keith Watson for the Nature Conservancy Council. No less than about fifty species we thought could well be extinct were found in the 1980s to be still present in the Glasgow area, sometimes at their old localities, sometimes at new places. A few examples are Crowberry, Field Mouse-ear, Lamb's Lettuce, Marjoram, Oak Fern and Water Chickweed.

Possil Marsh is the only legally fully protected nature reserve in the Glasgow area. It is also one of several Sites of Special Scientific Interest. An SSSI is a legal designation applied to land of special nature conservation interest by the Nature Conservancy Council, the statutory body for nature conservation in Britain. Owners and occupiers must give the NCC notice in writing if they intend to carry out, or cause or permit to be carried out, any operation listed in the notification by the NCC as likely to damage the special interest of the site. The NCC is obliged by law to take into account all operations likely to damage the flora, fauna or geological features by reason of which the land is of special interest. For botanical reasons alone, many of the places discussed in this book are worthy of some form of protection if they do not already have it. Add to these botanical reasons the interest of birds and other animals as well as geology and the case for any particular place may be greatly strengthened. Readers can help by joining the Scottish Wildlife Trust which is devoted to conservation and the Glasgow Natural History Society which deals with all aspects of natural history and has members expert in the identification of plants and animals.

Banks of the Clyde and Lochs

The straight, mostly vertical-sided stretch of the Clyde downstream of the tidal weir is far from devoid of botanical interest as Garden Angelica, Common Reed and a few salt-tolerant plants such as Sea Aster and Greater and Lesser Sea-spurreys show. However, the stretch upstream of the weir is much the more important: the banks support a varied and colourful flora of both natives and aliens. Some formal protection is provided by the Bothwell Castle Grounds Site of Special Scientific Interest, the northern boundary of which reaches close to Uddingston. However, all the way to Carmyle or even beyond, the natural history interest is such that conservation measures are very desirable. All the lochs in the Glasgow area have very considerable natural history and conservation interest, including those just outside, Woodend and Lochend Lochs. St Germains, Kilmardinny and Dougalston Lochs are very distinguished by the presence of Hybrid Yellow Waterlily. Cowbane adds even more to the lustre of St Germains Loch. None of these three lochs, all in Bearsden and Milngavie District, has any formal conservation protection. Of all the lochs only Bishop's Loch has been declared a Site of Special Scientific Interest. Even the reservoirs are very far from lacking in botanical interest. Mudwort at Balgray, Lesser Waterpepper at Glanderston, Trifid Bur-marigold at Waulkmillglen and Alpine Pondweed and Lesser Marshwort at Highflat forcefully make the point.

Canals, Railways and Motorways

Though long having lost its original function, the canal has recently seen a revival of its fortunes in its usefulness for leisure purposes and it is being upgraded by the British Waterways Board. Part of its importance to the residents of the Glasgow area is its natural history. There is a varied flora and fauna throughout its length, both along the sides and in the water. In the Glasgow area some plants are to be found exclusively or very largely in the canal. The Pinkston Basin in Port Dundas is important both to anglers who are more or less constantly present and to botanically minded natural historians. No part of the canal has as yet any formal protection for nature conservation.

Stretches of disused railway, both rural and urban, can be of great interest to the natural historian. The sinuous track between East Kilbride and Blantyre is a very good case. Both the west and east ends support well-developed woods and there are scrubby cuttings and embankments in the central portion with some flooded trackbed. The tree colonists include Ash, Beech, Birch and Wych Elm. Woodland ground flora includes Barren Strawberry, Dog's Mercury, Lesser Celandine, Lord's-and-Ladies, Opposite-leaved Golden Saxifrage, Polypody, Red Campion, Water Avens, Woodruff, Wood Sage and Wood Sorrel. Plants of marshy ground include Common Cottongrass, Common Sedge, Marsh Arrowgrass and Marsh Cinquefoil.

North-east-facing stonework of a former bridge has a profusion of Hart's-tongue with much commoner plants such as Cock's-foot, Male-fern and Rosebay Willowherb.

Strongly constrasting with this rural line because of a very urban setting, there is a small area of flooded sidings where an assemblage of plants perhaps without close parallel in Britain has developed. The flora includes Common Spotted and Northern Marsh Orchids, Brown Sedge, Lemon-scented Fern, Michaelmas Daisy, Prickly Heath, Rhododendron, Royal Fern, Round-leaved Sundew and Stag's-horn Clubmoss. Bay Willow is one of the colonising tall shrubs. In the Glasgow area no stretch of disused line has as yet been given any formal protection. Further details can be found in *Report on the Flora of Glasgow: Current Status With Reference to Conservation* which Keith Watson and I wrote for the Nature Conservancy Council.

Rubbly Wasteground

Though this type of habitat can accumulate a rich flora, it is not easy to conserve. Succession is bound to take place, given enough time, and the most interesting plants will disappear naturally as the space is all occupied and shade prevails. Only intermittent disturbance to produce patches of bare ground will keep the succession at the very early stage with the variety of plants, discussed in Chapter 7. In any case, such ground, especially if near the city centre, is very valuable commercially and will sooner or later be redeveloped. Perhaps a few very small areas of little or no interest to developers could be left for educational purposes to demonstrate succession, which often reaches a stage of scrubby Goat Willow as can be seen in many places in and around the city.

Bings and Coups

The unfamiliar weeds that can be found around coups deserve some attention, though the vast majority are unsuited to our climate and consequently are transient. Very rarely, some such alien may become established. However the rapidly changing and finally covered-over coups are of no interest to the conservationist. This is emphatically not the case with bings.

Bings are often unsightly, being conspicuous, incongruous, dark grey or reddish features of the landscape. They may long remain very visible and unsoftened by greenery because many plants may be slow colonists. Often close to large centres of population and surrounded by industrial dereliction, bings can be dangerous because they are unstable or they may be on fire internally and be releasing unpleasant fumes. No uncompacted spoil heaps have been produced for more than twenty years. In the past merely inconvenient waste products, bings are now valuable, the spoil being used for infilling of derelict industrial sites, road metal, the extraction of more coal, and even brick-making. For all these reasons, bings are a vanishing feature not just of the central Scottish landscape but of all mining areas.

Not all bings should be removed or landscaped and reclaimed (Dickson 1991). On the grounds alone that bings are part of the industrial history of Scotland, a few should be considered monuments, if not yet very ancient. Left undisturbed over decades, by natural colonisation and succession some bings develop woodland consisting not just of Birches and Willows but other trees, such as Oak and Wych Elm as some examples in the Glasgow area show. There may be plants noteworthy for their ecology, geography or evolution as those examples from the Glasgow area discussed above demonstrate. There are noteworthy plants found on bings elsewhere in Scotland. Fir Clubmoss, Alpine Clubmoss and the rare Moss *Buxbaumia aphylla* are a few examples.

Golf Courses and Heaths

The large number of courses and the substantial extent of ground they cover, with the diversity of vegetation, ensure importance for conservation. Apart from the examples discussed in Chapter 9, which make the courses at Whitecraigs and Cathkin Braes of great interest, there are other plants found in the Glasgow area only or mainly on golf courses. Common Cow-wheat (Windyhill), Field Mouse-ear and Bird's-foot (Sandyhills), Great Burnet (Lethamhill), Hybrid Fescue and Mountain Pansy (Kirkhill), Lemon-scented Fern (Ruchill and Clydebank). The heathy and rocky ground matters greatly to some of these and other plants. The great importance of British heaths on a European scale can be appreciated by reading *Ecology of Heathlands* by the Aberdeen University professor Charles Gimingham. There is now a Golf Course Wildlife Trust which tries to work with clubs to secure the survival of the special plants and animals. Only a little outside our area, Milngavie Golf Club has a special suite of plants inhabiting a little piece of marshy ground; there are Orchids including Early Marsh Orchid, one of the six extinct in the Glasgow area proper, Round-leaved Sundew, Common Butterwort, Lesser Clubmoss, Globeflower and Few-flowered Clubrush; the club has been co-operating with members of the Glasgow Natural History Society in trying to maintain this remarkable assemblage.

Cemeteries and Churchyards

From the botanical standpoint the Necropolis is by far the most interesting of the large cemeteries. Management continuing in the present way will allow the survival of such rare plants as Stag's-horn Clubmoss, Awl-leaved Pearlwort and *Hieracium strumosum*. As regards churchyards, if some patches of lawn were left uncut well into summer colourful meadows might be the result.

Gardens

Some of the cornfield weeds that are no longer to be found in the wild, or are much reduced there, are handsome plants, which can make colourful annuals for the garden. Good examples are Corncockle, Cornflower and Corn Marigold. Some at least can be bought as packets of wild flower seeds. That is the way to preserve these vanished plants. They are unlikely to become troublesome. Such is not the case with Welsh Poppy, with its beautiful yellow flowers, easily bought as seed; around Glasgow in the gardens of older houses such as my own in Milngavie it spreads freely. The drab cornfield weed, Field Woundwort, very rare in the Glasgow area now, has inhabited my small vegetable garden for at least twenty years without special favour or enmity from me and has never become a pest.

Woodlands

As elsewhere in Britain, even the most natural woodlands of the Glasgow area have been much altered by man's activities. These at the best semi-natural woods have been very greatly reduced, as is the case with heaths and wetlands. What little remains of these types of vegetation should not be thoughtlessly destroyed. If the trees have been shaped by ancient woodsmanship then the interest of the woodlands is enhanced; this is also so if the trees were planted over old rigs.

Some Glasgow area woods have a degree of formal protection. The Bothwell Woods (in part), Wilderness Plantation and the Cart/Kittoch Woods are all Sites of Special Scientific Interest. Kenmuir Wood, the locality of many uncommon and rare plants, has no protection nor does Bull Wood, nor Garscadden Wood.

Wetlands

Though it may be white with Cottongrasses in early summer and purple with Heather in late summer, a peat-bog is not an especially colourful place seen from the surrounding dry land. Bared stretches of peat and vertical sides of diggings are dark brown. The wet softness and water-filled ditches do not invite the too wary or uninitiated who will miss a range of subtle greens, reds, yellows, browns and greys displayed only at close hand by Bogmosses, Sundews, Bog Asphodel, Deergrass and Lichens. This drabness for most of the year to the eyes of the distant viewer belies the very special natural history interest of peat-bogs and leads some to think that wetlands are useless places, better drained and cut and perhaps even turned into coups. Perhaps the peat-bogs best known to the general public because of publicity in newspapers and on television are those in the flow country of the very far north and on the southern Hebridean Island of Islay. Though David Bellamy left Islay in the face of the anger of the islanders fearful of job losses, he was fundamentally right in trying to save the disputed peat-bog.

Throughout the British Isles and in lowland areas particularly, there is very little undamaged wetland. For natural history, scientific study and educational purposes some peat-bogs should be conserved, especially, but by no means only, those that are least damaged. Though the Nature Conservancy Council and the Scottish Wildlife Trust have given protection to some wetlands in various parts of Scotland, none of the peat-bogs in the Glasgow area are nature reserves. Those wetlands with deep peat at Bishopbriggs, Lenzie, Garnkirk-Gartloch and Commonhead are greatly altered remnants. Nevertheless they retain plants of much interest, not just those discussed above but others such as the semi-aquatic Bog Pondweed, the nitrogen-fixing Bog-myrtle and the nationally rare and decreasing Bog-Rosemary. Cut over and damaged in other ways as they are, these peat-bogs can recover by the growth of Bogmosses if only slowly. The recovery can be assisted by management that prevents further drying out. The extensive fens at Possil Marsh and Bishop Loch have some protection as SSSIs; there are other fen areas worthy of consideration for conservation.

Conclusion

Over the last 200 years Glasgow has spread very greatly from a small to a very large industrial city and the heavy industry has declined away over the last fifty years. The many activities of inhabitants have led to the extinction of about 100 species of plants, both natives and long-established aliens. At first this figure may seem very large but on reflection may be considered not too much of a disaster. Most extinctions took place last century when little concern was felt or expressed and indeed little could have been done by natural historians. In the very late twentieth century much has changed, not least with regard to conservation. However, losses may still happen if conservationists drop their guard. Another loss almost happened in 1989. The nationally rare Horsetail called Dutch Rush, growing as a very small stand in Kenmuir Wood, a fragment of old woodland, only just escaped extermination: unauthorised gravel extraction destroying parts of the wood came within ten metres of the plant. The loss of about 100 species is sad. Perhaps there is a kind of counterbalance in the many alien species some of which are conspicuous and colourful that have taken advantage of man-made habitats: Butterfly-bush, Fig, Giant Hogweed, Garden Angelica, Oxford Ragwort, Viper's Bugloss and many others. These invaders have no little appeal to the urban natural historian.

Change is inevitable in cities, which are fundamentally for people and not for wildlife. However, this is not to say that cities and flourishing wildlife are mutually exclusive. A love of natural history is one of the major characteristics of the human race, in Britain as much as if not more than anywhere. Stimulated by the magnificence of natural history and ecological films on television, more and more people find the enjoyment of plants and animals within or outwith cities to be a very important aspect of their lives. In the Environmental Age the human and wildlife inhabitants of cities and the surrounding countryside can be reconciled for the benefit of both.

BIBLIOGRAPHY

Allen, D. 1969. *The Victorian Fern Craze*. Hutchinson, London.

Boney, A D. 1988. *The Lost Gardens of Glasgow University*. Christopher Helm, London.

Burton, R M. 1983. *Flora of the London Area*. London Natural History Society.

Curl, J S. 1974. *The Cemeteries and Burial Grounds of Glasgow*. Glasgow District Council.

Dickson, Camilla and Dickson, James. 1988. 'The diet of the Roman Army in Deforested Central Scotland'. *Plants Today* 1, 121–126.

Dickson, J H. 1990. '*Epipactis helleborine* in Gardens and Other Urban Habitats. *Urban Ecology*, ed H Sukopp, 105–9. SPB Publishing, The Hague.

Dickson, J H. 1991. 'Conservation and the Botany of Bings'. *Transactions of the Botanical Society of Edinburgh* 45, 493–500.

Dickson, J H and Gauld, W. 1988. 'Mark Jameson's physic plants a gynaecological garden in 16th century Glasgow?' *Scottish Medical Journal* 32, 60–62.

Dickson, J H and Watson, K. 1988. *Report on the Flora of Glasgow: Current Status With Reference to Conservation*. Nature Conservancy Council.

Dickson, J H and Watson, K. 1989. *A Red Data List of Native and Long Known Alien Plants in the Glasgow Area*. Nature Conservancy Council.

Gibb, A. 1983. *Glasgow: The Making of a City*. Chapman and Hall, London.

Gilbert, O L. 1989. *The Ecology of Urban Habitats*. Chapman and Hall, London.

Gilbert-Carter, H. 1955. *Glossary of the British Flora*. Cambridge University Press, 2nd edition.

Gimingham, C H. 1972. *Ecology of Heathlands*. Chapman and Hall, London.

Glasgow Naturalist. This, the annual journal of the Glasgow Natural History Society, has published articles on the plants of the Glasgow area for many years.

Goode, D. 1986. *Wild in London*. Michael Joseph, London.

Goodyear, J. *The Greek Herbal of Dioscorides*. Oxford.

Grierson, R. 1931. 'Clyde Casuals'. *Glasgow Naturalist* 9, 1–51.

Grierson, S. 1986. *The Colour Cauldron*. Mrs S Grierson, Newmiln Farm, Perth.

Grigson, G. 1958. *The Englishman's Flora*. Phoenix House.

Grigson, G. 1974. *A Dictionary of Plant Names*. Allen Lane, London.

Hammerton, D. 1990. 'Cause for Concern?' *Discover Scotland The Sunday Mail Guide to Scotland's Countryside.* Page 1373, volume 4, part 49.

Headrick, J. 1808. 'Biographical sketch of the late Rev. David Ure'. *Scots Magazine* LXX, 903–5.

Hennedy, R. 1865. *The Clydesdale Flora; a Description of the Flowering Plants and Ferns of the Clyde District.* David Robertson, Glasgow. Fifth and last edition, 1891.

Hopkirk, T. 1813. *Flora Glottiana. A Catalogue of the Indigenous Plants on the Banks of the River Clyde and in the Neighbourhood of the City of Glasgow.* John Smith, Glasgow.

Hutchings, M J. 1989. 'Conservation and the British Orchid Flora'. *Plants Today* 2, 50–8.

Jackson, P W and Skeffington, M S. 1984. *The Flora of Inner Dublin.* Royal Dublin Society.

Jardine, W G. 1990. Drumlins. Page 1367, *Discover Scotland The Sunday Mail Guide to Scotland's Countryside*, volume 4, part 49.

King, T and Boyd, D A. 1893. 'Report on the Disappearance of Native Plants'. *Proceedings and Transactions of the Natural History Society of Glasgow* IV, 44–8.

Lee, J R. 1933. *The Flora of the Clyde Area.* John Smith, Glasgow.

Lee, J R. 1953. Additions to the flora of the Clyde area. *Glasgow Naturalist* 17, 65–82.

Lindsay, Maurice. 1987. *Victorian and Edwardian Glasgow from Old Photographs.* B T Batsford, London.

Loudon, J C. 1843. *On the Laying Out, Planting and Managing of Cemeteries and on the Improvement of Churchyards.* Longman, London.

Mabberley, D J. 1987. *The Plant-book.* Cambridge University Press.

MacDonald, H. 1854. *Rambles Round Glasgow, Descriptive, Historical and Traditional.* Thomas Murray, Glasgow. The 1910 Edition, revised by Rev G H Morrison, discusses changes since MacDonald's time.

Macpherson, P. 1982. 'The Doctrine of Signatures.' *Glasgow Naturalist* 20, 191–210.

Nicholson, Max. 1987. *The New Environmental Age.* Cambridge University Press.

Oakley, C A. 1990. *The Second City.* Blackie, Glasgow, 4th edition.

Odum, S. 1978. *Dormant Seeds in Danish Ruderal Soils.* Horsholm Arboretum, Denmark.

Page, C. 1982. *The Ferns of Britain and Northern Ireland.* Cambridge University Press.

Page, C. 1988. *Ferns.* Collins, London.

Patton, D and Rennie, W. 1955. The Plants of Possil Marsh. *Glasgow Naturalist* 17, 161–72.

Perring, F H and Farrell, L. 1983. *British Red Data Book: 1 Vascular Plants.* The Royal Society for Nature Conservation, Lincoln.

Perring, F H and Walters, S M. 1976. *Atlas of the British Flora.* EP Publishing, Wakefield.

Price, R. 1989. *Scotland's Golf Courses.* Aberdeen University Press.

Rackham, Oliver. 1986. *The History of the Countryside.* J M Dent, London.

Rae, J H. 1988. *City Profile. Facts and Figures about Glasgow, 1988.* Planning Department, City of Glasgow.

Simpson, W D. 1985. *Bothwell Castle*, 3rd edition revised by D J Breeze and J R Hume. HMSO, Edinburgh.

Stearn, W T and Campbell, E. 1986. *The European Garden Flora*, ed Walters, S M et al, vol 1, 233–46. Cambridge University Press.
Stewart, N. 1987. 'Rare Orchid protection and depredation'. *Plant Press* 2, 1.
Strang, J. 1831. *Necropolis Glasguensis*. Atkinson, Glasgow.
Thomas, B. 1981. *The Evolution of Plants and Flowers*. Peter Lowe.
Ure, D. 1793. *The History of Rutherglen and East Kilbride*. David Niven, Glasgow.
Walker, Agnes, Millar, Jean, Nove, Irene and Jarvis, Michael. 1988. Plants from the Kenmuir District of the Clyde 1815–1987. *Glasgow Naturalist* 21, 375–99.
Zohary, M. 1982. *Plants of the Bible*. Cambridge University Press.

Further Reading and Action

Field Guides
Any good bookshop has a range of field guides large enough to be almost perplexing. The reader should be aware that there is no guide specifically written for the identification of the wild plants of western Scotland. Often writers of field guides are not Scottish and consequently have little experience of Scotland and they have written on a British scale or even a continental scale; such guides can be misleading in their too great scope.

A book that is authoritative with good text, illustrations and keys for identification is *The Wild Flower Key* (1981, recently reprinted, Frederick Warne). The author is Francis Rose, a well-known botanist, though even he thinks that Mudwort is extinct in Scotland; see pages 57, 78 and 79 of this book.

The Wild Flowers of the British Isles (1983, Macmillan) written by David Streeter, has beautiful, accurate paintings by Ian Garrard. Except for a very few, the plants are all reproduced at life size though the book does not say so. Grasses, Sedges, Rushes and trees are not covered. Sadly the book was remaindered not long after publication but is worth trying to find.

The Macmillan Field Guide to British Wildflowers (1989, Macmillan) was written by the very experienced Franklyn Perring and Max Walters with photographs by Andrew Gagg. Again trees, Grasses, Rushes and Sedges are not included nor are rarities. Any page of photographs shows several plants all at differing sizes. You must read the sizes given in the text.

Action
The three societies listed here all hold indoor meetings, publish journals and run field excursions led by expert botanists.

Anyone wishing to develop an interest in the wild plants of the Glasgow area and the west of Scotland should join the *Glasgow Natural History Society* by writing to the General Secretary at Kelvingrove Museum, Glasgow G3 8AG.

The Botanical Society of Scotland (until recently the Botanical Society of Edinburgh) deals with many aspects of botany and is strongly interested in conservation of plants in Scotland. A key to Scottish Orchids was published in 1991. Write to the General Secretary c/o the Royal Botanic Garden, Inverleith Row, Edinburgh EH3 5LR.

The Botanical Society of the British Isles is that society which publishes very helpful guides dealing with the identification of British plants which, like Sedges and Willows, can be difficult to recognise. Write to the General Secretary c/o Department of Botany, The Natural History Museum, Cromwell Road, London SW7 5BD.

INDEX

The English names of the plants are almost all taken from *The English Names of Wild Flowers* by Dony, J. H. *et al.* 1986. Second edition, Botanical Society of the British Isles. The few exceptions are mostly names I learned as a boy in Glasgow. After each English name I have inserted the scientific name with the authorities.

INDEX

INDEX